Fuzzy Computing in Data Science

Scrivener Publishing
100 Cummings Center, Suite 541J
Beverly, MA 01915-6106

Sustainable Computing and Optimization

Series Editor: Prasenjit Chatterjee, Morteza Yazdani and Dilbagh Panchal

Scope: The objective of "Sustainable Computing and Optimization" series is to bring together the global research scholars, experts, and scientists in the research areas of sustainable computing and optimization from all over the world to share their knowledge and experiences on current research achievements in these fields. The series aims to provide a golden opportunity for global research community to share their novel research results, findings, and innovations to a wide range of readers, present globally. Data is everywhere and continuing to grow massively, which has created a huge demand for qualified experts who can uncover valuable insights from data. The series will promote sustainable computing and optimization methodologies in order to solve real life problems mainly from engineering and management systems domains. The series will mainly focus on the real life problems, which can suitably be handled through these paradigms.

Publishers at Scrivener
Martin Scrivener (martin@scrivenerpublishing.com)
Phillip Carmical (pcarmical@scrivenerpublishing.com)

Fuzzy Computing in Data Science

Applications and Challenges

Edited by

Sachi Nandan Mohanty

Prasenjit Chatterjee

and

Bui Thanh Hung

Scrivener Publishing

WILEY

This edition first published 2023 by John Wiley & Sons, Inc., 111 River Street, Hoboken, NJ 07030, USA and Scrivener Publishing LLC, 100 Cummings Center, Suite 541J, Beverly, MA 01915, USA

For more information about Scrivener publications please visit www.scrivenerpublishing.com.

Wiley Global Headquarters
111 River Street, Hoboken, NJ 07030, USA

For details of our global editorial offices, customer services, and more information about Wiley products visit us at www.wiley.com.

Library of Congress Cataloging-in-Publication Data

ISBN 978-1-119-86492-9

Cover image: Pixabay.Com
Cover design by Russell Richardson

Set in size of 11pt and Minion Pro by Manila Typesetting Company, Makati, Philippines

Printed in the USA

10 9 8 7 6 5 4 3 2 1

Dedication

The Editors would like to dedicate this book to their parents, life partners, children, students, scholars, friends and colleagues.

Contents

Preface

The term fuzzy logic describes the concept of incomplete information. In real life, decision makers frequently encounter situations when it becomes very tough to apprehend whether a statement is true or false. Fuzzy logic allows a great deal of flexibility in reasoning at the time. The fuzzy algorithms aid in the solution of a problem and make the best feasible decision after taking into account all relevant data. They are mostly used in rule-based automatic controllers and based on the likelihood of an individual output being assigned to a particular input. Several organizations use advanced data analysis and business intelligence models to boost their performance. These analytical models entail working with data from a variety of sources. Fuzzy algorithms are an example of the transition from accuracy to imprecision. The present book aims to present how fuzzy computing can be explored to handle several challenges in real time data science applications. Machine learning approaches are used in computing environments to talk about intelligence, predictive and prescriptive features in understanding data analytics and the ways to deal them respectively. Fuzzy based systems are targeted to prophesy recommended improvements for better decision making in several domains and fields. Various robust traits of fuzzy systems paves its applications in Information Technology, Tourism, Education, E-Commerce, Agriculture, Health Care, Manufacturing, Sales, Retail and many more.

This book explains how to use various fuzzy-based models to solve real-time industrial challenges. The book is organized in 17 chapters. The inclusion of extensive guidance for using fuzzy-based models in each chapter is a unique feature of this book. The following is a chapter-by-chapter summary:

Chapter 1 discusses about efficient feature reduction and extraction with efficient search strategies to extract a set of spectral bands from a given imagery. In Chapter 2, a fuzzy-based framework is designed to detect and warn if a user is wearing a mask correctly or not when in public locations, using public services, or even while conversing with someone. This framework can be used in public places to monitor the public and alert users if they are not wearing their mask appropriately. Chapter 3 examines a fundamental FL framework and its applicability in the healthcare industry. A descriptive study was undertaken to look into and focus on the ranking, classification, data mining, feature selection, pattern recognition, and optimization processes in healthcare decision-making. Chapter 4 provides a structural overview and systematic review of the extensive research work that has been done on the use of fuzzy logic in various dimensions, as well as to assist researchers in gaining understanding into the issue. Chapter 5 presents a fuzzy decision making application for resource scheduling problems. Chapter 6 focuses on sales data analysis using four different types of seeds: spinach, black gramme, ginseng, and fodder sorghum. The proposed work used a fuzzy logic technique called clustering and artificial intelligence algorithms to evaluate sales of various seeds over time, as well as the performance of sales in a specific location. In Chapter 7, a single-pole (single-valued) decision making problem is solved using Shannon's entropy weight calculation integrated m-polar fuzzy ELECTRE-I methodology. For demonstration purposes, two industrial selection problems are considered: flexible production system selection and cutting fluid selection. The concept of a fuzzy set in decision-making is discussed in Chapter 8. This chapter provides an outline of how fuzzy set theory can be used to various decision-making issues. The purpose of Chapter 9 is to describe the psychological state of migrant workers in Covid-19. Migrant workers from Odisha's two southern districts (Khurdha and Gajapati) have been considered. In addition, this chapter discusses how fuzzy logic might aid in improving the psychological well-being of such migrants. Chapter 10 describes a fuzzy-based edge AI method for smart healthcare transformation for a better future. Chapter 11 attempts to provide a complete guideline for picking the best video-conferencing software for a higher education institution based on their needs, and then uses the Fuzzy- Technique for Order Preference by Similarity to Ideal Solution (TOPSIS) method to rank the various alternatives. Chapter 12 presents an application of Fuzzy Analytic Hierarchy Process (AHP) with extended Saaty's scale and Goal Programming model for a case study of Indian commercial bank. Chapter 13 investigates a new fuzzy-based mathematical approach for analysing and evaluating static workstation performance parameters in order to make accurate and correct decisions. In Chapter 14, a model for calculating the power created by ocean waves is described. Furthermore, by using a fuzzy optimizer, it can maximise the overall power generated. In Chapter 15, the m-polar fuzzy set TOPSIS approach is used to answer the challenge of selecting a non-traditional machining process (NTM). Chapter 16 attempts to select the optimal material handling devices by combining a number of traditional fuzzy group decision-making models. In Chapter 17, a theoretical study of critical chain project management (CCPM) is offered, and the practical application of CCPM is demonstrated.

The Editors
August 2022

Acknowledgement

The editors wish to express their warm thanks and deep appreciation to those who provided input, support, constructive suggestions, and comments and assisted in editing and proofreading of this book.

This book would not have been possible without the valuable scholarly contributions of authors across the globe.

The editors avow the endless support and motivation from their family members and friends.

The editors are very grateful to all the members who served in the editorial and review process of the book.

Mere words cannot express the editors' deep gratitude to the entire Scrivener Publishing team, particularly Mr. Martin Scrivener for keeping faith and showing the right path to accomplish this very timely book.

Finally, the editors use this opportunity to thank all the readers and expect that this book will continue to inspire and guide them for their future endeavour.

The Editors

1

Band Reduction of HSI Segmentation Using FCM

V. Saravana Kumar[1*], S. Anantha Sivaprakasam[2], E.R. Naganathan[3], Sunil Bhutada[1] and M. Kavitha[4]

[1]Department of IT, SreeNidhi Institute of Science and Technology, Hyderabad, India
[2]Department of CSE, Rajalakshmi Engineering College, Chennai, India
[3]Department of CSE, Koneru Lakshmaiah Education Foundation, Vaddeswaram, AP, India
[4]Research Scholar, Manonmaniam Sundaranar University, Tirunelveli, India

Abstract

Hyperspectral has carried hundreds of nonoverlapping spectral channels of a specified scene, clustering is one of the approaches for diminishing the size of these large data sets. Segmentation is intricate for the raw data; however, it is likely for the reduced band of HSI. To lessen the band size, the classical clustering methods for example K-means, Fuzzy C-means are accomplished. An integrated image segmentation procedure built on interband clustering and intraband clustering is proposed. The interband clustering is performed by K-means clustering and Fuzzy C-means clustering algorithms, despite the fact the intraband clustering is executed using particle swarm segmentation (PSO) clustering algorithm. The performance of the K-means algorithm is subject to initial cluster centers. Besides, the final partition should be contingent on the initial configuration. The clustering consequences have profoundly been subject to the number of clusters stated. It is essential to provide refined direction for defining the number of clusters with the purpose of attaining appropriate clustering consequences. Davies Bouldin (DB) index is one of the reliable methods to outline the number of clusters for these clustering methods. The hyperspectral bands are clustered, and a band which has extreme variance from each cluster is preferred. This tactic forms the diminished set of bands. PSC (EEOC) accomplishes the segmentation process on the reduced bands. In conclusion,

**Corresponding author*: drsaravanakavi@gmail.com

Sachi Nandan Mohanty, Prasenjit Chatterjee and Bui Thanh Hung (eds.) Fuzzy Computing in Data Science: Applications and Challenges, (1–20) © 2023 Scrivener Publishing LLC

there is a comparison of the result produced for K-means worked with EEOC and FCM worked with EEOC in various HSI scenes.

Keywords: K-means, Fuzzy C-means, band reduction, PSO, cluster, centroid

1.1 Introduction

Hyperspectral [14] has carried hundreds of nonoverlapping spectral channels [11] of a given scene, clustering [13, 21] is one of the methods for reducing the size of these large data sets. Despite displaying the size of hyperspectral scene [17], the device does not support to display the scene directly. Segmentation is complicated for the raw data, whereas it is possible for the reduced band scene. Even though hyperspectral data [3] can provide finely resolved details about the spectral properties [4] to be identified, it also has some limitation. When dealing with such high-dimensional data [24, 25], one is faced with the "curse of dimensionality" problem One popular way to tackle the curse of dimensionality [7] is to employ a feature extraction technique [22]. To diminish the band size, the classical clustering methods [30], such as K-means, Fuzzy C-means [27] are handled. The former can be sensitive to the initial centers, while the results from the latter depend on the initial weights. Here, an integrated image segmentation [2] process based on interband clustering and intraband clustering is proposed. The interband clustering is performed by K-means clustering or Fuzzy C-means clustering algorithms, whereas the intraband clustering is executed using particle swarm segmentation (PSO) clustering algorithm. The performance of the K-means algorithm depends on initial cluster centers. Besides, the final partition depends on the initial configuration. The clustering results have heavily depended on the number of clusters specified. It is necessary to provide educated guidance for determining the number of clusters in order to achieve appropriate clustering results. Davies Bouldin (DB) index is one of the reliable methods to determine the number of clusters for these clustering methods. The hyperspectral bands [18] are clustered and a band [34], which has maximum variance, from each cluster is chosen. This forms the reduced set of bands. PSC (EEOC) [29] performs the segmentation process on the reduced bands. Finally, the result produced from K-means worked with EEOC and FCM worked with EEOC in various HSI [19] scenes was compared.

1.2 Existing Method

1.2.1 K-Means Clustering Method

Theoretically, K-means [1] is a typical algorithm. Now that it is elementary and expeditious, it is attractive in practice. To begin with, it segregates the input dataset into K- lusters. Each cluster is described by an adaptively changing centroid, starting from some initial values named seed points. K-means enumerates the squared distances between the inputs and centroids, and assigns inputs to the nearest centroid. The procedure continues until there is no significant change in the location of class mean vectors between successive iteration of the algorithms. Apparently, the performance of the K-means algorithm depends on initial cluster centers, whereas the final partition depends on the initial configuration.

1.2.2 Fuzzy C-Means

In Fuzzy C-means clustering [8], data elements can belong to more than one cluster and associated with each element is a set of membership levels. FCM clustering [9] is a process of assigning these membership levels, and then using them to assign data elements to one or more clusters. The aim of FCM [10] is to minimize an objective function. Contrary to traditional clustering analysis methods, which distribute each object to a unique group, fuzzy clustering algorithms gain membership values between 0 and 1 that indicate the degree of membership for each object to each group. Obviously, the sum of the membership values for each object to all the groups is definitely equal to 1. Different membership values show the probability of each object to different groups.

The main limitation of the FCM algorithm [16] is its sensitivity to noise. The FCM algorithm implements the clustering task for a data set by minimizing an objective-function subject to the probabilistic constraint that the summation of all the membership degrees of every data point to all clusters must be one. This constraint results in the problem of this membership assignment, that noise is treated the same as points close to the cluster centers. However, in reality, these points should be assigned very low or even zero membership in either cluster.

Like K-means, the clustering results have heavily depended on the number specified. It is also necessary to provide an educated guidance for determining the number cluster in order to achieve appropriate clustering results. Davies Bouldin (DB) index is one of the reliable methods to determine the number of clusters for these clustering methods.

1.2.3 Davies Bouldin Index

Davies Bouldin index was introduced in 1979 by David L. Davies and Donald W. Bouldin. It is one of the methods for evaluating clustering algorithms. This is an internal evaluation scheme, where the validation of how well the clustering has done using quantities and features inherent to the dataset.

Many other distance metrics can be used, in the case of manifolds and higher dimensional data, where the Euclidean distance may not be the best measure for determining the clusters. It is important to note that this distance metric has to match with the metric used in the clustering scheme itself for meaningful results.

Davies-Bouldin (DB) index is dependent both on the data, as well as the algorithm. The Davies–Bouldin index measures the average of similarity between each cluster and its most similar one. As the clusters have to be compact and separated the lower Davies-Bouldin index produces a better cluster configuration.

In this work, the interband clustering and intraband clustering approach has proposed. In interband clustering part; existing method such as K-means and Fuzzy C-means method are applied to reduce the band size. In intraband clustering part; a lightweight algorithm, namely Enhanced Estimation of Centroid (EEOC) [33] is proposed. This method is examined with the abovementioned clustering method by applying in various hyperspectral scenes.

1.2.4 Data Set Description of HSI

The dataset contains a variety of hyperspectral remote sensing [15], which are acquired from airborne and satellite. In this work, certain data, such as Salinas A, Salinas Valley, Indian Pines, Pavia University, and Pavia Centre, are handled.

The original scene and its corresponding ground truth image are downloaded from the link. http://www.ehu.eus/ccwintco/index.p... Hyperspectral_Remote_Sensing_Scenes

These scenes are a widely used benchmark for testing the accuracy of hyperspectral data [20] classification [23, 26] and segmentation [31].

1.3 Proposed Method

1.3.1 Hyperspectral Image Segmentation Using Enhanced Estimation of Centroid

These topics explain the integration of intraband and interband cluster approach for segmentation of hyperspectral image in a synergistic fashion. It demonstrates the advantage of the advanced properties of both analysis techniques in combined fashion of clustering method. The ultimate goal is to improve the analysis and interpretation of hyperspectral image. Owing that, classifications [5, 12], as well as segmentation [6], face problems related to the extremely high dimensionality [32] of the hyperspectral images are dominated by mixed pixels and un-mixing techniques are crucial for a correct interpretation and exploitation of the data. This chapter explored the integration of intraband and interband clustering methods for segmentation process with the ultimate goal of obtaining more accurate methods for the analysis of hyperspectral scene without increasing significantly the computational complexity of the process. There are many things to do for achieving this goal; to begin with, to improve unsupervised segmentation [28] by learning a task-relevant measure of spectral similarity from the feature matrix approaches. Besides, interband and intraband cluster techniques for segmentation are proposed. In addition, various existing clustering techniques are applied to interband part. Moreover, for improving the accuracy of various clustering methods, Davies-Bouldin Index is used to determine the number of clusters.

Furthermore, for intraband cluster, a new and novel algorithm entitled as Enhanced Estimation of Centroid (EEOC), which is the modification of the particle swarm clustering method, is proposed. The modification is that the particle has updated their positions and computed the distance matrix only once per iteration. In addition, performances about these methods are evaluated in different scenarios. Performance measurements are used in this research to investigate the efficacy of this system.

The working principle of hyperspectral image segmentation using EEOC is depicted in Figure 1.1. To begin with, the hyperspectral scene in the form of mat file is read; the feature matrix could be constructed based on mean or median absolute deviation (MAD), standard deviation (STD), variance (VAR). In addition, apply one of the clustering processes, namely K-means or Fuzzy C-means. Since these clustering method works depend on the number of cluster, Davies Bouldin (DB) index is used to determine the number of clusters. Furthermore, the dimensional reduction process could be carried out by this clustering method by picking out one band

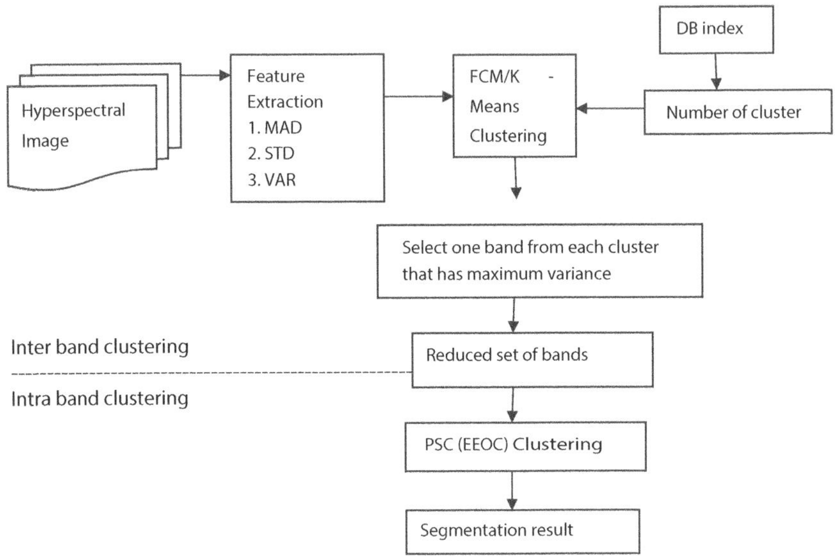

Figure 1.1 Flowchart for hyperspectral image segmentation using EEOC.

from each cluster, such as 204 bands of hyperspectral image are reduced to below twenty bands, i.e., one band is pick out based on maximum variance from each cluster. In particle swarm clustering algorithm, the input is reduced set of bands from K-means or FCM clustering algorithm. PSC (EEOC) is used for segmentation on the reduced bands.

1.3.2 Band Reduction Using K-Means Algorithm

K-means is an iterative technique that is used to partition the hyperspectral feature matrix into K clusters. It is one of the simplest unsupervised learning algorithms to solve the well-known clustering problem. The aim of this method is to minimize the sum of squared distances between all points and the cluster center.

The K-means algorithm is reduced the hyperspectral feature matrix is shown in Figure 1.2. Get the hyperspectral feature matrix and the number of clusters (determined from DB Index). Choose the K initial cluster centers, after which the new cluster is computed, 204 bands of the hyperspectral matrix are reduced to 17 bands, i.e., one band is selected according to the maximum variance from each cluster. Finally, the reduced set of bands is obtained.

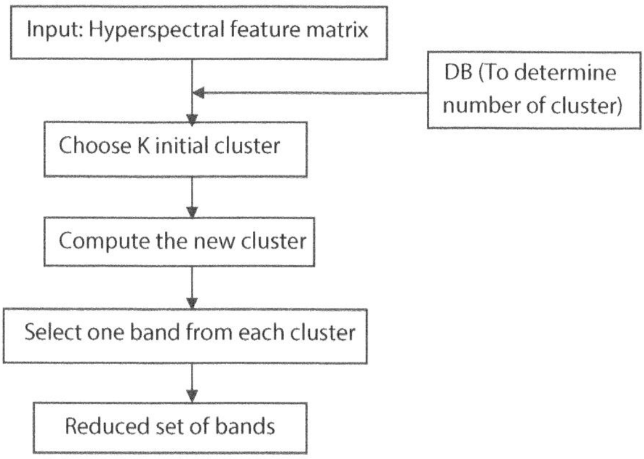

Figure 1.2 Band reduction using K-means algorithm.

1.3.3 Band Reduction Using Fuzzy C-Means

Fuzzy C-means is an iterative technique that is used to partition the hyperspectral feature matrix into membership levels. Data elements can belong to more than one cluster in this method and associated with each element in a set of membership levels. It is a process of assigning these membership levels and then using them to assign data elements to one or more clusters.

The Fuzzy C-means algorithm is reduced the hyperspectral feature matrix is as abovementioned Figure 1.3. Get the hyperspectral feature matrix and the number of clusters that determine the DB Index. The membership matrix is initialized after which fuzzy cluster is calculated. Then, membership matrix is updated. 204 bands of the hyperspectral matrix are reduced to 17 bands i.e. one band is selected according to the maximum variance from each cluster. Finally, the reduced set of bands is obtained. The above flowchart shows the flow of band reduction using Fuzzy C-means clustering algorithm.

By using this above said clustering method, the hyperspectral bands are reduced into below 20 bands, depending on the scene. The DB index helps to determine the number of clusters, based on this number of cluster the bands are reduced, i.e., applying this clustering method into the hyperspectral feature matrix. These matrices are clusters that depend on the number of clusters. Then, pick out one band from each cluster, which has a maximum variance, i.e., number of clusters is 18, that implies the maximum variance band is pick out from each cluster and thus this hyperspectral

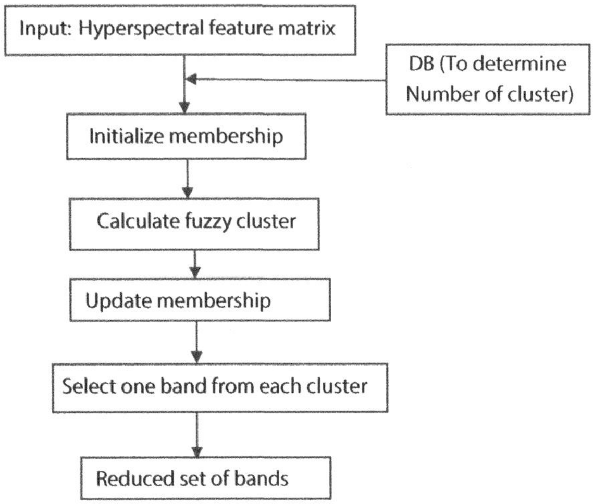

Figure 1.3 Band reduction using Fuzzy C-means.

data are reduced into 18 bands. This reduced band is allowed to segmentation, which is carried out by a light weight clustering algorithm. This algorithm is entitled as enhanced estimation of centroid (EEOC), which is the modification of the particle swarm clustering method performed. The modification is that the particle has updated their positions and computed the distance matrix only once per iteration.

1.4 Experimental Results

The performance of EEOC algorithm is analyzed and experimented. Its performance is noticed to be satisfactory in term of reduction in the time complexity and the efficiency of each position update.

1.4.1 DB Index Graph

Figure 1.4(a) depicts DB Index Graph. In x-axis, number of clusters (i.e., the value is incremented by 1 pixel) and in y-axis, DB value (i.e., the value is increment by 0.05 pixels) is determined.

For Salinas_A scene, the DB-value is least at 17th cluster. So, the DB Index of this scene is considering as 17. For Indian Pines the DB value is least at 11th cluster. So, the DB Index is considered as 11 for this scene. For Salinas Valley Scene, the DB value is least at 10th cluster. So, the DB Index

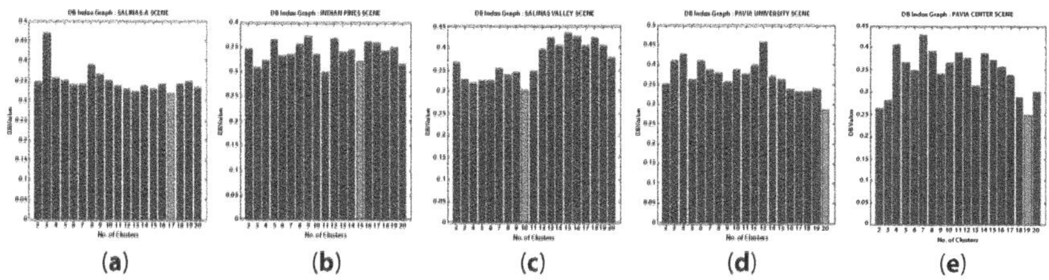

Figure 1.4 DB Index Graph. (a) Salina_A scene (b) Indian Pines (c) Salinas Valley (d) Pavia University and (e) Pavia Center scene.

is considered as 10 for this scene. For Pavia University, the DB value is least at 20th cluster. So, the DB Index is considered as 20 for this scene. For Pavia Centre Scene, the DB value is least at 19th cluster. So, the DB Index is considered as 19 for this scene.

1.4.2 K-Means–Based PSC (EEOC)

Figure 1.5(a) shows the segmentation result for the hyperspectral scene, namely Salinas_A, is processed by K-means based PSC (EEOC). In the interband clustering, K-means is utilized to reduce the band, whereas PSC (EEOC) is performed in intracluster. For this reason this approach is named as K-means based PSC (EEOC). Figures 1.6(a), 1.7(a), 1.8(a), 1.9(a) portrays the segmentation results for the hyperspectral scenes, namely Indian Pines, Salinas Valley, Pavia University, and Pavia Center, respectively. The pixels, which are clustered based on K-means working with PSC (EEOC), are displayed in Table 1.1.

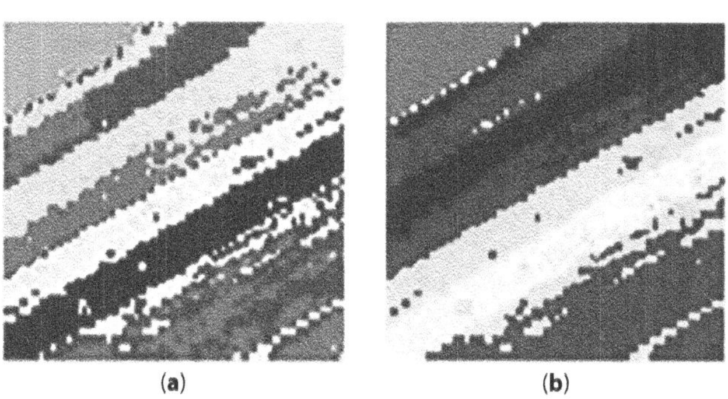

Figure 1.5 Results for Salinas_A, (a) K-means + PSC (EEOC) (b) FCM + PSC (EEOC).

Table 1.1 Pixels clustered based on PSC (EEOC) with K-means and FCM for Salinas_A.

SALINAS _A 83 × 86 × 204		K-Means + PSC (EEOC)	FCM + PSC (EEOC)
	Clusters	**Pixels**	**Pixels**
	1	1284	1371
	2	1492	2788
	3	1081	507
	4	611	1
	5	507	1431
	6	2163	1040

1.4.3 Fuzzy C-Means–Based PSC (EEOC)

Figures 1.5(b), 1.6(b), 1.7(b), 1.8(b), and 1.9(b) portray the segmentation results for FCM based on PSC (EEOC), which is examined on this HSI. The pixels are clustered based on FCM +PSC (EEOC) is listed in Table 1.1.

The following figures (Figures 1.5 to 1.9), left side images (a images), depict the results for K-means working with PSC (EEOC) of the various hyperspectral scenes. The right side listed images (b images) portray the results for FCM working with PSC (EEOC).

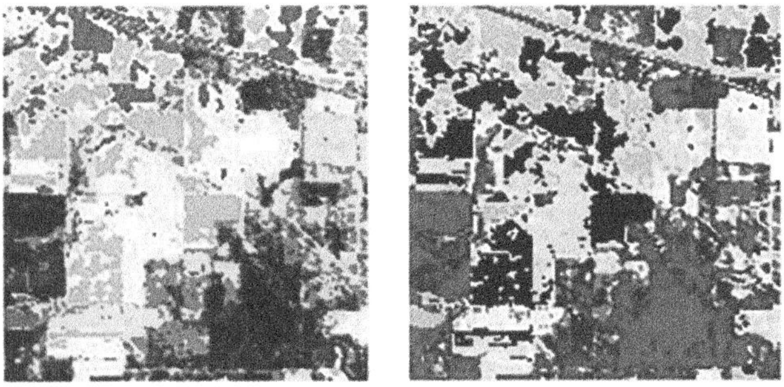

Figure 1.6 Results for Indian Pines (a) K-means + PSC (EEOC) (b) FCM +PSC (EEOC).

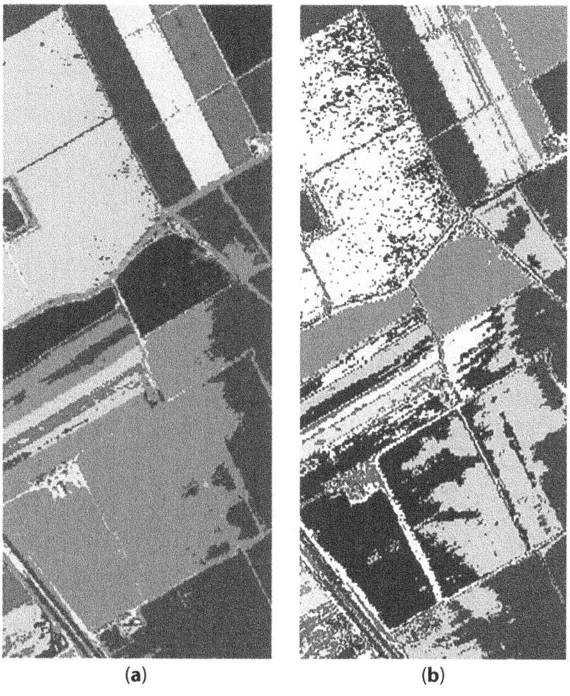

Figure 1.7 Results for Salinas Valley. (a) K-means + PSC (EEOC) (b) FCM + PSC (EEOC).

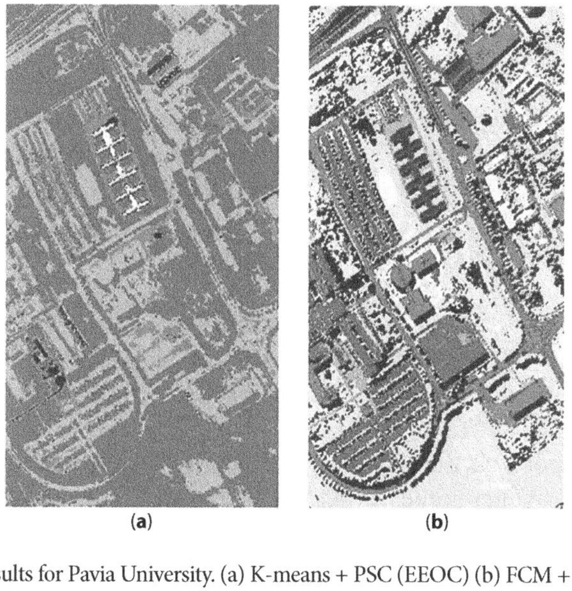

Figure 1.8 Results for Pavia University. (a) K-means + PSC (EEOC) (b) FCM + PSC (EEOC).

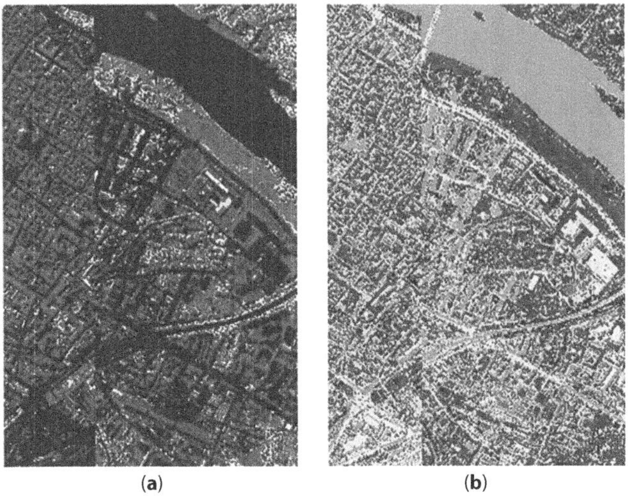

Figure 1.9 Results for Pavia Centre. (a) K-means + PSC (EEOC) (b) FCM + PSC (EEOC).

1.5 Analysis of Results

The following tables illustrate the resultant of the number of pixels, which are clustered. The number of the cluster is defined based on the ground truth, as well as important features of the abovementioned HSI.

Table 1.1 depicts the result of performance based on pixel wise cluster for the Salinas_A scene. As per the reference of the ground truth image, Salinas_A scene has segmented with six clusters. The size of this scene is $83 \times 86 \times 204$. The 204 bands are reduced into 17 bands as per the Davies-Bouldin Index and finally this 83 x 86 pixels are clustered into six, such as 7138 pixels are grouped into these clusters. From Table 1.1, K-means working with PSC (EEOC) clustered the pixels equally, whereas 4th cluster has one pixel for FCM.

Table 1.2 portrays the result of performance based on pixel wise cluster for the Indian Pines scene. As per the reference of the ground truth image, the Indian Pines scene has segmented with seven clusters. The size of this scene is $145 \times 145 \times 200$. The 200 bands are reduced into 11 bands as per the Davies-Bouldin Index and finally this 145×145 pixels are clustered into seven, such as 21,025 pixels are grouped into these clusters.

Table 1.3 portrays the result of performance based on pixel wise cluster for the Salinas Valley scene. As per the reference of the ground truth image, this scene has segmented with seven clusters. The size of this scene is

512 × 217 × 204. The 204 bands are reduced into 10 bands as per the Davies-Bouldin Index and finally this 512 × 217 pixels are clustered into seven, and 111,104 pixels are grouped into these clusters.

Table 1.2 Pixels clustered based on PSC (EEOC) with K-means and FCM for Indian Pines.

Indian Pines 145 × 145 × 200		**K-Means + PSC (EEOC)**	**FCM + PSC (EEOC)**
	Clusters	**Pixels**	**Pixels**
	1	2035	3354
	2	2953	3522
	3	1393	3700
	4	3177	1497
	5	2662	2193
	6	3597	3603
	7	5208	3156

Table 1.3 Pixels clustered based on PSC (EEOC) with K-means and FCM for Salinas Valley.

Salinas Valley 512 × 217 × 204		**K-Means + PSC (EEOC)**	**FCM + PSC (EEOC)**
	Clusters	**Pixels**	**Pixels**
	1	9006	22894
	2	29424	21034
	3	6695	5021
	4	35776	11939
	5	23797	18292
	6	6328	7784
	7	78	24140

Table 1.4 portrays the result of performance based on pixel wise cluster for the Pavia University scene. As per the reference of the ground truth image, this scene has segmented with nine clusters. The size of this scene is 610 × 340 × 103. The 103 bands are reduced into 20 bands as per the DB-Index, and finally these 610 × 340 pixels are clustered into nine, such as 207,400 pixels are grouped into these clusters.

Table 1.5 portrays the results of performance based on pixel wise cluster for the Pavia Centre scene. As per the reference of the ground truth image, Pavia Centre scene has segmented with nine clusters. The size of this scene is 1096 × 715 × 102. These 102 bands are reduced into 17 clusters, as per the Davies-Bouldin Index and finally this 1096 × 715 pixels are clustered into six, such as 783640 pixels are grouped into these clusters.

Table 1.6 depicted the time taken for execution of these methods. These approaches are applying on the above said hyperspectral scenes. From the table, K-means + PSC method takes minimum time to execute for all HSI, whereas FCM + PSC method takes more time for processed.

Table 1.4 Pixels clustered based on PSC (EEOC) with K-means and FCM for Pavia University.

Pavia University 610 × 340 × 103		**K-Means + PSC (EEOC)**	**FCM + PSC (EEOC)**
	Clusters	**Pixels**	**Pixels**
	1	2176	104374
	2	656	4899
	3	637	162
	4	116561	23482
	5	13031	1013
	6	72789	53200
	7	128	2150
	8	610	2769
	9	812	15351

Table 1.5 Pixels clustered based on PSC (EEOC) with K-means and FCM for Pavia Centre.

Pavia Center 1096 × 715 × 102		**K -Means + PSC (EEOC)**	**FCM + PSC (EEOC)**
	Clusters	**Pixels**	**Pixels**
	1	454182	38,965
	2	22	156,770
	3	52534	88,386
	4	26	63,542
	5	130076	194,464
	6	9294	13,021
	7	132714	115,264
	8	20	113,205
	9	4772	23

Table 1.6 Elapsed time in seconds for PSC (EEOC) with K-means and FCM.

Input	**Size**	**K-means + PSC (EEOC)**	**FCM + PSC (EEOC)**
SALINAS_A	83 × 86 × 204	85.4219s	102.5313s
INDIAN PINES	145 × 145 × 200	95.6563s	104.6719s
SALINAS VALLEY	512 × 217 × 204	277.8438s	283.7031s
PAVIA UNIVERSITY	610 × 340 × 103	9.98e+02s	10.48e+2s
PAVIA CENTER	1096 × 715 × 102	19.68e+2s	18.11e+2s

Since, in these works, particle swarm clustering method is used, fitness value is one of the parameters to measure the accuracy. The least fitness value indicates the optimum results.

Table 1.7 depicts that the fitness value of EEOC worked with K-means/FCM. PSC (EEOC) worked with FCM produce least fitness value other than the Salinas_A scene. This scene majorly contains the straight line.

Table 1.7 Fitness value for PSC (EEOC) with K-means and FCM.

Input	K-means + PSC (EEOC)	FCM + PSC (EEOC)
SALINAS_A	15.99e+5	17.96e+05
INDIAN PINES	96.17e+5	92.11e+05
SALINAS VALLEY	52.38e+6	40.82+06
PAVIA UNIVERSITY	174.56e+6	73.65e+06
PAVIA CENTER	34.16e+07	88.72e+07

For this reason, EEOC worked with K-means produce optimum results. In addition, here, it is to be noted that the fitness value for K-means + PSC (EEOC) is a nearby triple of FCM + PSC (EEOC) for Pavia University. For Pavia Centre, the fitness value of FCM + PSC (EEOC) is more than double of K-means + PSC (EEOC). As a whole, EEOC is producing optimum results when worked with the abovementioned clustering methods.

1.6 Conclusions

An integrated image segmentation process based on interband clustering and intraband clustering is proposed. EEOC is used for segmentation process on the reduced bands. The hyperspectral bands are clustered, and a band which has a maximum variance from each cluster is chosen. K-means and FCM are used for interband clustering part, i.e., reduced the band. The Davies-Bouldin Index is used to determine the number of clusters. Finally, reduced set of bands are obtained. PSC (EEOC) takes the reduced set of bands as input and produces the segmentation result.

To be crisp, FCM-based PSC (EEOC) has shown promising results when compared to K-means–based PSC (EEOC) for Salinas Valley, Indian Pines, and Pavia university scenes. Despite producing higher accuracy, FCM clustering takes more time for computation. K-means + PSC (EEOC) clustering produced close results to FCM + PSC (EEOC) clustering for Salinas Valley and Indian Pines. For Pavia University, crystal clearly FCM + PSC (EEOC) produces better results in terms of fitness value and accuracy. As a result, EEOC shows best performance when worked with K-means and Fuzzy C-means clustering method.

References

1. Abel Guilhermino da, S., Filho, A.C.F. *et al.*, Hyperspectral images clustering on reconfigurable hardware using the K-means algorithm, in: *Integrated circuit and system design*, IEEE, US, 2003.
2. Sivaprakasam, A., Naganathan, E.R., Saravanakumar, V., Wavelet based cervical image segmentation using morphological and statistical operations. *J. Adv. Res. Dyn. Control Syst.*, 10-03, 1138–1147, 2018.
3. Benhalouche, F.Z., Karoui, M.S., Deville, Y., Boukerch, I., Ouamr, A., Multi-sharpening hyperspectral remote sensing data by multiplicative joint-criterion linear-quadratic nonnegative matrix factorization. *2017 IEEE International Workshop of Electronics, Control, Measurement, Signals and their Application to Mechatronics (ECMSM)*, Donostia-San Sebastian, 2017, pp. 1–6, 2017.
4. Song, B., Li, J., Mura, M.D. *et al.*, Remotely sensed image classification using sparse representations of morphological attribute profiles. *IEEE Trans. Geosci. Remote Sens.*, 52, 5122–5136, 2014.
5. Bouziani, M. *et al.*, Rule-based classification of a very high resolution image in an urban environment using multispectral segmentation guided by cartographic data. *IEEE Trans. Geosci. Remote Sens.*, 48, 3198–3211, 2010.
6. Chen, T.W. *et al.*, Fast image segmentation based on K-means clustering with histograms in HSV color space. *Multimedia Signal Processing, IEEE 10th Workshop on*, pp. 322–325, 20082008.
7. Honga, D., Yokoya, N., Chanussot, J., Xu, J., Zhu, X.X., Learning to propagate labels on graphs: An iterative multitask regression framework for semi-supervised hype-rspectral dimensionality reduction. *ISPRS J. Photogramm. Remote Sens.*, 158, 35–49, 2019.
8. Li., D.-F., *Linear programming models and methods of matrix games with payoffs of triangular fuzzy numbers*, Springer, Berlin, Heidelberg, 2016.
9. Li, D.-F., *Decision and game theory in management with intuitionistic fuzzy sets*, Springer, Heidelberg, Germany, 2014.
10. Fan, J. *et al.*, Single point iterative weighted fuzzy C-means clustering algorithm for remote sensing image segmentation. *Pattern Recognit.*, 42, 2527–2540, 2009.
11. Ferraris, V., Dobigeon, N., Wei, Q., Chaber, M. *et al.*, Detecting changes between optical images of different spatial and spectral resolutions: A fusion-based approach. *IEEE Trans. Geosci. Remote Sens.*, 56, 3, 1566– 1578, 2018.
12. Gao Yan, J., Mas B, F., Maathuis, H.P., Xiangmin, Z., Van Dijk, P.M., *Comparison of pixel-based and object-oriented image classification approaches—A case study in a coal fire area*", Wuda, Inner Mongolia, China. *Int. J. Remote Sens.*, 27, 18, 4039–4055, 20 September 2006.
13. Ghosh, A. *et al.*, "*Fuzzy clustering algorithms for unsupervised change detection in remote sensing images*". *Inform. Sci.*, 181, 699–715, 2011.

14. Govender, M., Chetty, K., Bulcock, H., A review of hyperspectral remote sensing and its application in vegetation and water resource studies. *Water SA*, 33, 145–151, 2007.
15. Guo, B. *et al.*, Customizing kernel functions for SVM-based hyperspectral image classification. *Image Process. IEEE Trans.*, 17, 622–629, 2008.
16. An, J.-J. and Li, D.-F., A linear programming approach to solve constrained bi-matrix games with intuitionistic fuzzy payoffs. *Int. J. Fuzzy Syst.*, 21, 3, 908–915, 2019.
17. Kerekes, J.P. and Baum, J.E., Hyperspectral imaging system modeling. *Linc. Lab. J.*, 14, 117 –130, 2003.
18. Benediktsson, J.A., Palmason, J.A., Sveinsson, J.R., Classification of hyperspectral data from urban areas based on extended morphological profiles. *IEEE Trans. Geosci. Remote Sens.*, 43, 3, 480–491, 2005.
19. Bioucas-Dias, J.M. and Plaza, A., Hyperspectral remote sensing data analysis and future challenges. *IEEE Geosci. Remote Sens. Mag.*, 1, 2, 6–36, June 2013.
20. Jung., A., Kardevan, P., Tokei, L., Detection of urban effect on vegetation in a less built-up Hungarian city by hyperspectral remote sensing. *Phys. Chem. Earth*, 30, 255–259, 2005.
21. Kavitha, M. *et al.*, Enhanced clustering technique for segmentation on dermoscopic images. *IEEE - 4th International Conference on Intelligent Computing and Control Systems (ICICCS 2020)*, Madurai, TN, pp. 956–961, 2020May 2020.
22. Li, Z., Nie, F., Chang, X., Yang, Y., Zhang, C., Sebe, N., Dynamic affinity graph construction for spectral clustering using multiple features. *IEEE Trans. Neural Netw. Learn. Syst.*, 29, 12, 6323–6332, 2018.
23. He, L., Chen, X., Li, J., Xie, X., Multiscale super pixel wise locality preserving projection for hyperspectral image classification. *MDPI Appl. Sci.*, 27, 2161, 1-21, May 2019.
24. Liu, L., Nie, F., Wiliem, A., Li, Z., Zhang, T., Lovell, B.C., Multi-modal joint clustering with application for unsupervised attribute discovery. *IEEE Trans. Image Process.*, 27, 9, 4345–4356, 2018.
25. Khodadadzadeh, M., Li, A.P., Ghassemian, H., Bioucas -Dias, J.M., Li, X., Spectral–spatial classification of hyperspectral data using local and global probabilities for mixed pixel characterization. *IEEE Trans. Geosci. Remote Sens.*, 52, 10, 6298–6314, 2014.
26. Pooja, K., Nidamanuri, R.R., Mishra, D., Multi-scale dilated residual convolutional neural network for hyperspectral image classification. *10th Workshop on Hyperspectral Imaging and Signal Processing: Evolution in Remote Sensing (WHISPERS)*, Amsterdam, Netherlands, pp. 1–5, 2019.
27. Saravanakumar, V., Kavitha, M. *et al.*, Fuzzy C-means technique for band reduction and segmentation of hyperspectral satellite image. *Int. J. Fuzzy Syst. Appl. (IJFSA)*, 10, Article 5, 4, 79–100, 2021Sep 2021.
28. Saravanakumar, V., Kavitha, M. *et al.*, Robust K-means technique for band reduction of hyperspectral image segmentation, in: *Application of hybrid*

metaheuristic algorithms for image processing, vol. 890, pp. 81–104, Springer - Studies In Computational Intelligence, Switzerland AG, 2020.

29. Saravanakumar, V. and Naganathan, E.R., Hyperspectral image segmentation based on enhanced estimation of centroid with fast k-means. *Int. Arab J. Inform. Technol. Jordon*, 15, 5, 904–911, 2018.
30. Saravana Kumar., V., Kavitha, M. *et al.*, Segmentation of hyperspectral satellite image based on classical clustering method. *Int. J. Pure Appl. Math.*, 118, 9, 813–820, 20182018.
31. SaravanaKumar., V. and Naganathan, E.R., Segmentation of hyperspectral image using JSEG based on unsupervised clustering algorithms. *ICTACT J. Image Video Process.*, 06, 02, 1152–1158, 2015.
32. SaravanaKumar., V. and Naganathan., E.R., A survey of hyperspectral image segmentation techniques for multiband reduction. *AJBAS*, 9, 7, 226 - 451, 2015.
33. SaravanaKumar., V., Naganathan., E.R. *et al.*, *Multiband Image Segmentation by using Enhanced Estimation of Centroid (EEOC). Int. Interdiscip. J., Tokyo, Japan*, 17- A, 1965–1980, July 2014.
34. Yang, H. *et al.*, Unsupervised hyperspectral band selection using graphics processing units. *IEEE J. Appl. Earth Observ. Remote Sens.*, 4, 660–668, 2011.

2

A Fuzzy Approach to Face Mask Detection

Vatsal Mishra, Tavish Awasthi, Subham Kashyap, Minerva Brahma, Monideepa Roy* and Sujoy Datta

School of Computer Engineering, KIIT Deemed to be University, Bhubaneswar, India

Abstract

With the severe outbreak of the pandemic of COVID-19, several guidelines were laid down by the CDC and WHO for the effective containment of the spread. Some of these measures were proper wearing of adequate masks, social distancing, sanitation, frequent washing of hands, various degrees of lockdown and isolation. Here, we have designed a fuzzy based framework for the automated detection to detect and warn if a user is wearing a mask correctly or not, (e.g., covering both mouth and nose) when in public places, using public services or even while talking to someone. This framework can be implemented at public places to monitor the public and warn a user if he/she is not correctly wearing the mask. The architecture used for the object detection purpose is Single Shot Detector (SSD) because of its good performance accuracy and high speed. Transfer learning in neural networks has also been used to finally find out the presence or absence of a face mask in an image or a video stream. Experimental results show that our model performs well on the test data with 100% and 99% precision and recall, respectively.

***Keywords*:** COVID-19, pandemic, fuzzy decision making, mask wearing

**Corresponding author*: monideepa.roy@gmail.com

Sachi Nandan Mohanty, Prasenjit Chatterjee and Bui Thanh Hung (eds.) Fuzzy Computing in Data Science: Applications and Challenges, (21–36) © 2023 Scrivener Publishing LLC

2.1 Introduction

With the outbreak of severe acute respiratory syndrome coronavirus 2 (SARS-CoV-2) infection in Wuhan, Hubei, China in December 2019, it has been declared as pandemic by the World Health Organization (WHO) due to its global outbreak. After the worldwide breakout of the COVID-19 pandemic, the number of global new cases had already exceeded 70 million and the number of death was around a quarter of a million according to the information provided by the world health organization (WHO) (at the time of writing this chapter). Therefore, there was an urgent need to devise a means to restrict the spread of the disease to as much of an extent as possible. Guidelines were issued to people to curtail the spreading of COVID-19, which included wearing a mask covering the nose and mouth, maintaining a physical distance of six feet from others, avoiding crowds, and washing hands frequently.

This has compelled civil administration to enforce wearing face masks in public due to rising COVID-19 cases all over the world. Before the onset of the COVID-19 pandemic, common people usually wore masks for protection against air pollution. The main users of masks for medical or surgical purposes were doctors and practitioners in the healthcare domain. However, as scientists have now proven that the wearing face masks effectively impedes COVID-19 transmission, almost all countries have issued a health advisory to make the wearing of masks compulsory, by all people who are travelling outside their homes.

Over five million cases of infections by COVID-19 were registered in less than 6 months across 188 countries. According to scientists, the virus was found to be primarily airborne, which meant that it spreads through close proximities and in crowded places. In this chapter, we have presented a framework for a mask face detection model that is based on computer vision and deep learning. The proposed model detects whether a person is wearing a mask or not, and whether a mask has been correctly worn according to guidelines. This framework can also be integrated with a surveillance camera for face detections and by raising proper alerts that can effectively reduce the spread of the virus.

The model integrates deep learning and classical machine learning techniques with OpenCV, TensorFlow, and Keras. Deep transfer learning has been used for the feature extractions.

Our main contributions are as follows:

1. Initially, the main target was to build a large data set consisting of masked faces, properly wearing masked face detection

data set and then include nonmasked face data set. Few predefined classes were included regarding the target cases.
2. To make the model a bit more rigid, it was adjusted to application scenarios. As shown in the result, the model when deployed was able to achieve 99% accuracy.
3. OpenCV-Python have been used for the real time detection of a face mask on a person, without the usage of much resources. The model is also capable of detecting faces in different orientations and can also detect faces, which are occluded with a good deal of accuracy.

2.2 Existing Work

Device detection or object detection is one of the most fundamentally challenging tasks in the computer vision and is being greatly applied in the applications of face detection, digit detection, handwriting detection, pedestrian detection, etc. In the area of object detection, face detection has been in high demand. Due to the outbreak of the global pandemic (COVID-19), face detection has achieved a lot of attention and research. At present, various types of machine learning algorithms have been developed for enhanced object detection or face detection [7, 9].

CNN-Based Object Detectors

Although deep learning was first proposed in 2006 , it did not receive much attention until 2012. With the advent of a large amount of labeled datasets and the increase of computational power, deep learning has been found to be very effective in extracting intrinsic structure and high-level features. A lot of progress has been made to solve different types of problems in AI, and especially in the areas where the data is multidimensional and the feature are difficult to hand-engineer. Some examples are speech recognition, natural language processing, and computer vision. Deep learning is also very useful in areas, like business analysis, medical diagnostics, art creation and image translation.

Of these areas, computer vision is one field that has had remarkably successful applications of deep learning. This is mainly because since vision is the most important sense in terms of navigation and recognition in human beings, it provides the maximum information about the surroundings. With the rapid development of image sensors, computer vision [4] is now applied to a wide spectrum of applications like autonomous driving, visual surveillance, facial recognition, etc.

The fundamental step for many computer vision applications is object detection. The first step to extract the spatial and classification information of an object is usually done by convolutions neural networks (CNNs) as a first step. After that more models are added for specific applications. So most of the research on computer research will depend on the performance of object detection. Convolutional Neural Network is a kind of artificial neural network used for image recognition and processing that is especially designed to gather pixel data. It is a fundamental example of deep learning, which offer a more advanced evolution of artificial intelligence to systems that simulate different types of biological brain activities.

Face-Mask Detection

In view of the COVID 19 outbreak, it has become compulsory for all persons to wear masks when they are in public places or where there are clusters of persons. Since not everyone is following this mandate properly, it has become necessary to monitor the detection of masks and whether the masks were being used properly to ensure the safety of people. Apart from the people who are suspected of being infected with COVID-19 needing to wear masks to prevent the infection from spreading, healthy people also need to wear masks to protect themselves from infection. When these face masks are fitted properly, they can effectively reduce the transmission of infection by obstructing the parties expelled from a cough or sneeze. But unfortunately, the effectiveness of face masks are reduced mostly due to improper wearing. Therefore, it is very important to develop an automatic mask wearing detection approach, which can detect the presence of a mask and whether it has been worn properly. In Wang *et al.* [5], the authors have generated a new anchor level attention algorithm for obstructed face recognition that might boost the different regions of the face and improve the accuracy by combining the anchor point allocation strategy and data expansion. They have also assembled different types of datasets for this type of studies.

Now, a drawback of that was that in real time performance the different areas were compressed and so the resulting accuracy of the given dataset was just 89.6%. In Loey *et al.* [3], the authors have inherited the hybrid transfer learning model and machine learning methods for a nicer classification. The final accuracy reached 99.64% on the real time face dataset (RMFD). In Cabani *et al.* and Chiang *et al.* [1, 2], the authors have proposed masked face images based on facial feature landmarks and created a huge dataset of 137,016 masked images that gave us more data for training. Along with this, they have created mobile applications that made people wear masks correctly.

In Qin and Li [6], the authors have proposed a new facemark wearing condition identification method, which combines image super resolution and classification networks (SRCNet), based on unconstrained 2D facial images.

Our proposed method uses a customized deep learning model to identify if a person is wearing a mask correctly or not.

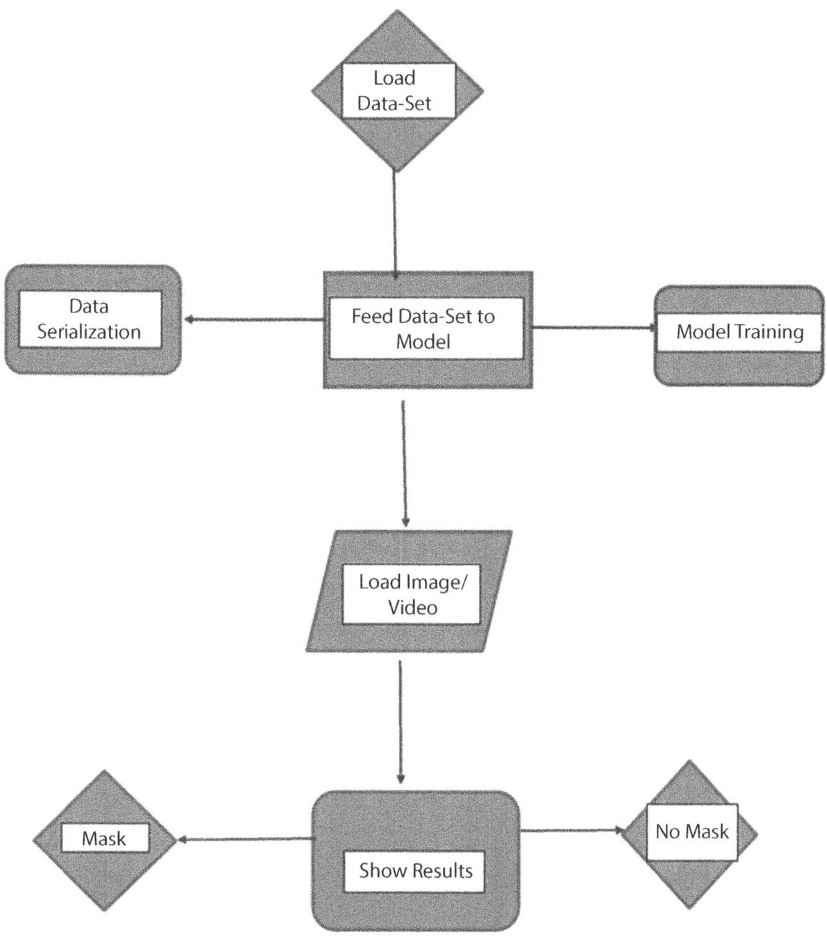

Figure 2.1 The proposed framework.

2.3 The Proposed Framework

It has already been proven that proper wearing of masks and face coverings can prevent the wearer from contracting or transmitting the COVID-19 virus to a large extent. However, it has been observed that many people do not wear their masks properly or not wear masks at all. The following framework (Figure 2.1) was designed to train a custom deep learning model to detect whether a person is or is not wearing a mask.

2.4 Set-Up and Libraries Used

In this section an introduction of the various libraries and their function are given along with the setup details. Some of the libraries used were TensorFlow, Keras, Matplotlib (Figure 2.2), etc.

Libraries and Their Functions

TensorFlow

TensorFlow has been created by the Google Brain team and is an open source library for numerical computation and large-scale machine learning. TensorFlow is a collection of machine learning and deep learning (aka neural networking) models and algorithms and makes these algorithms useful as a common metaphor. Tensorflow provides multiple application programming interfaces (APIs), which can be classified as low-level API and high-level API.

```
# import the necessary packages
from tensorflow.keras.preprocessing.image import ImageDataGenerator
from tensorflow.keras.applications import MobileNetV2
from tensorflow.keras.layers import AveragePooling2D
from tensorflow.keras.layers import Dropout
from tensorflow.keras.layers import Flatten
from tensorflow.keras.layers import Dense
from tensorflow.keras.layers import Input
from tensorflow.keras.models import Model
from tensorflow.keras.optimizers import Adam
from tensorflow.keras.applications.mobilenet_v2 import preprocess_input
from tensorflow.keras.preprocessing.image import img_to_array
from tensorflow.keras.preprocessing.image import load_img
from tensorflow.keras.utils import to_categorical
from sklearn.preprocessing import LabelBinarizer
from sklearn.model_selection import train_test_split
from sklearn.metrics import classification_report
from imutils import paths
import matplotlib.pyplot as plt
import numpy as np
import os
```

Figure 2.2 Some of the libraries used.

Keras
Keras is the high-level API of TensorFlow 2. It is a highly productive interface for solving machine learning problems, with a focus on modern deep learning. It contains essential abstractions and building blocks useful for the development and shipping of machine learning solutions with high iteration velocity. It is also relatively easy to learn and work with because it has a Python frontend with a high level of abstraction while having the option of multiple back-ends for computation purposes. While his makes it a bit slower than other deep learning frameworks, it is very suitable for new learners.
Imutils
This package includes a series of OpenCV + convenience functions that perform basics tasks such as translation, rotation, resizing, and skeletonization.
Numpy
NumPy is a Python library which is used for working with arrays. It also contains necessary the functions needed for linear algebra, Fourier transform, and matrices.
Open CV-Python:
OpenCV-Python is the Python API of OpenCV. It combines the best qualities of OpenCV C++ API and Python language. So OpenCV-Python is an appropriate tool for fast prototyping of computer vision problems.
Matplotlib
Matplotlib [8] is the visualization library in Python for 2D plots of arrays. Matplotlib is a multi-platform data visualization library built on NumPy arrays. The main advantage of visualization is that it allows for the visual access to huge amounts of data in easily comprehensible visuals. Matplotlib consists of several plots, like line, bar, scatter, histogram, etc.
Scipy:
The Scipy package contains various toolboxes dedicated for common issues in scientific computing. There are different submodules for different applications, such as interpolation, integration, optimization, image processing, statistics, special functions, etc.

2.5 Implementation

In this section, we present the implementation for the creation of a detector that can differentiate between faces with mask and without masks. In order to create this model, we need some amount of data in the form of images. So we have a dataset containing 3833 images of faces with and

```
DIRECTORY = r"C:\Users\vatsa\Documents\Face-Mask-Detection-master\Face-Mask-Detection-master\dataset"
CATEGORIES = ["with_mask", "without_mask"]

# grab the list of images in our dataset directory, then initialize
# the list of data (i.e., images) and class images
print("[INFO] loading images...")

data = []
labels = []

for category in CATEGORIES:
    path = os.path.join(DIRECTORY, category)
    for img in os.listdir(path):
        img_path = os.path.join(path, img)
        image = load_img(img_path, target_size=(224, 224))
        image = img_to_array(image)
        image = preprocess_input(image)

        data.append(image)
        labels.append(category)
```

Figure 2.3 The dataset and directory.

without masks (Figure 2.3). Now in order to get the model trained using this dataset, first we need to import all the necessary libraries, which were mentioned above in the previous section. After this, we move to the training phase by reading all the images and assigning them some list and the paths associated with these images and then labeling them in the same way.

Now the images are contained in two separate folders with_mask and without_mask. So it is easier to get the label by extracting the name from the path. After this, we move forward to data augmentation which helps to increase the variety of data available for training models. The next step is to compile the model and train it on the data. After our model is trained, we will be plotting a graph (Training Loss Graph) to monitor the learning curve.

```
[INFO] compiling model...
[INFO] training head...
Epoch 1/20
95/95 [==============================] - 122s 1s/step - loss: 0.6138 - accuracy: 0.7053 - val_loss: 0.1493 - val_accuracy:
0.9870
Epoch 2/20
95/95 [==============================] - 105s 1s/step - loss: 0.1855 - accuracy: 0.9551 - val_loss: 0.0803 - val_accuracy:
0.9909
Epoch 3/20
95/95 [==============================] - 115s 1s/step - loss: 0.1198 - accuracy: 0.9676 - val_loss: 0.0560 - val_accuracy:
0.9922
Epoch 4/20
95/95 [==============================] - 106s 1s/step - loss: 0.0822 - accuracy: 0.9801 - val_loss: 0.0459 - val_accuracy:
0.9922
Epoch 5/20
95/95 [==============================] - 102s 1s/step - loss: 0.0695 - accuracy: 0.9808 - val_loss: 0.0408 - val_accuracy:
0.9922
Epoch 6/20
95/95 [==============================] - 111s 1s/step - loss: 0.0694 - accuracy: 0.9825 - val_loss: 0.0378 - val_accuracy:
0.9922
Epoch 7/20
95/95 [==============================] - 117s 1s/step - loss: 0.0550 - accuracy: 0.9851 - val_loss: 0.0381 - val_accuracy:
0.9896
Epoch 8/20
95/95 [==============================] - 107s 1s/step - loss: 0.0523 - accuracy: 0.9820 - val_loss: 0.0345 - val_accuracy:
```

Figure 2.4 Training the model.

```
95/95 [==============================] - 109s 1s/step - loss: 0.0327 - accuracy: 0.9934 - val_loss: 0.0293 - val_accuracy:
0.9909
Epoch 15/20
95/95 [==============================] - 108s 1s/step - loss: 0.0330 - accuracy: 0.9912 - val_loss: 0.0273 - val_accuracy:
0.9935
Epoch 16/20
95/95 [==============================] - 102s 1s/step - loss: 0.0322 - accuracy: 0.9924 - val_loss: 0.0399 - val_accuracy:
0.9844
Epoch 17/20
95/95 [==============================] - 104s 1s/step - loss: 0.0360 - accuracy: 0.9917 - val_loss: 0.0303 - val_accuracy:
0.9935
Epoch 18/20
95/95 [==============================] - 106s 1s/step - loss: 0.0266 - accuracy: 0.9914 - val_loss: 0.0280 - val_accuracy:
0.9935
Epoch 19/20
95/95 [==============================] - 104s 1s/step - loss: 0.0235 - accuracy: 0.9920 - val_loss: 0.0269 - val_accuracy:
0.9935
Epoch 20/20
95/95 [==============================] - 113s 1s/step - loss: 0.0210 - accuracy: 0.9959 - val_loss: 0.0273 - val_accuracy:
0.9935
[INFO] evaluating network...
```

Figure 2.5 Training the model.

The dataset we are working on consists of 3833 images, with 1915 images containing images of people wearing masks and 1918 images with people without masks.

Steps for the training model:

With this, we are representing the training phase of the model where we try to train 3833 datasets of masked faces and unmasked faces, which is being trained in batches of 20 (Figures 2.4, 2.5).

2.6 Results and Analysis

Once the training job is done and complete, the next step is to extract the newly trained interface graph which will later be used to perform the object detection or for the face detection for whether the subject is wearing the mask properly or not. This will be done using TensorFlow object detection program. Figure 2.6 shows the stage of evaluation of the networks and the precision scores. Figure 2.7 shows the training loss and accuracy graph. Figures 2.8 through 2.13 show the results when the model was tested for various types of faces with masks and without masks.

```
[INFO] evaluating network...
              precision    recall  f1-score   support

   with_mask       0.99      0.99      0.99       383
without_mask       0.99      0.99      0.99       384

    accuracy                           0.99       767
   macro avg       0.99      0.99      0.99       767
weighted avg       0.99      0.99      0.99       767
```

Figure 2.6 Evaluation of the networks and the precision scores.

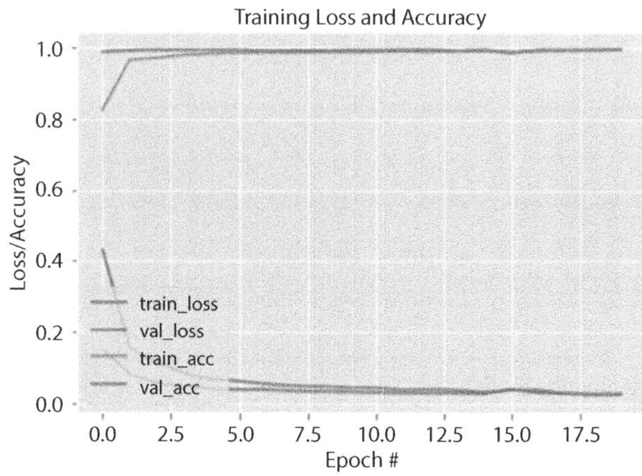

Figure 2.7 Training loss and accuracy graph on the COVID-19 dataset.

Figure 2.8 In real time with cap, face detected without a mask with 100% accuracy.

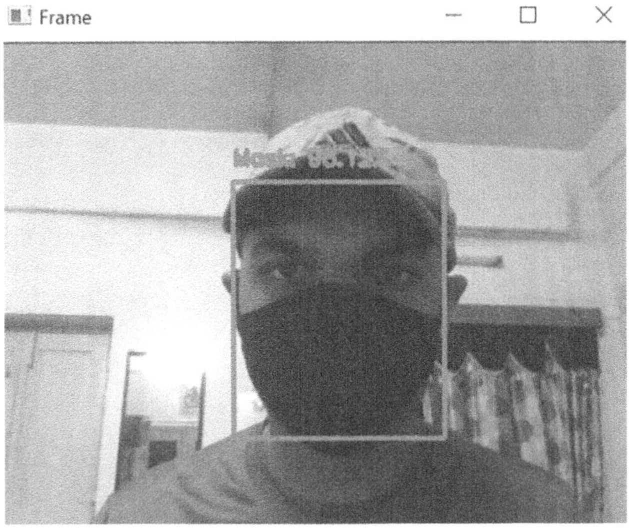

Figure 2.9 In real time, with cap, face detected with the mask with 99% accuracy.

Figure 2.10 In real time, the persons face in the image is detected without mask with 99% accuracy.

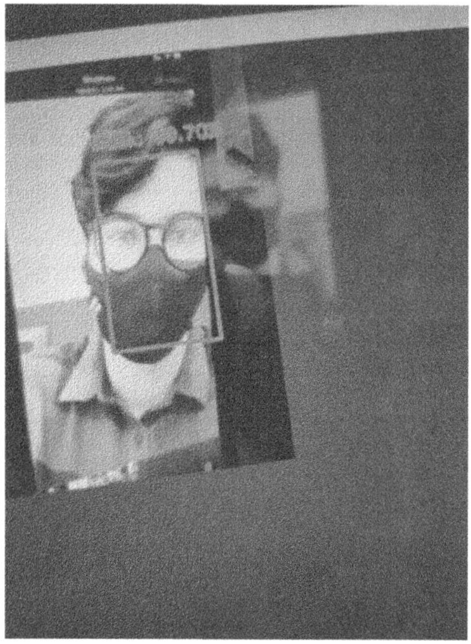

Figure 2.11 In real time the persons face in the image is detected with the mask with 99% accuracy.

Figure 2.12 In real time, the persons face in the image is detected without mask with 99% accuracy.

Figure 2.13 In real time ,the persons face in the image is detected with the mask with 92% accuracy.

2.7 Conclusion and Future Work

In this chapter, we developed a model for detection of masks on persons in real time. After the pandemic, more than half the world has made wearing of face masks a compulsory rule in order to avoid the spread of the virus. The main aim of our project was to expand it on existing flash drives, IP cameras, and CCTV cameras in order to detect whether people are wearing masks or not, and if they are wearing masks, whether the masks are being worn accurately or not. This detection mechanism can then be installed in different devices in order to get the live feed of the people entering the premises wearing mask or not. Further, we can also add to trigger with a siren or a bell sound if someone enters not wearing a mask properly. This can be used for enhancing vision-based monitoring systems toward several applications, such as checking whether people are wearing masks as per stipulations or not, or generating crowd statistics.

The proposed model can be used as a surveillance tool at public places to check whether a person is wearing a mask properly or not according to stipulations. Here, MobileNet is used as the backbone of the architecture and is suitable for conditions of both low and high computations. For the extraction of more robust features, transfer learning has been used to adopt weights from a similar task of face detection, which was trained on a very large dataset. We have used OpenCV, TensorFlow, Keras, Scipy, matplotlib for detecting if a person is wearing a mask or not. The testing of the models were done with images, as well as real-time video streams. The accuracy of the model is being done by tuning the hyper parameters. By implementing the face mask detection framework, we can detect if the person is wearing a face mask or not and if it is being worn accurately or not, and based on this, whether the person should be allowed entry to the public places, which would be very useful in screening in places where large volumes of persons enter a public place.

The main limitation here, however, was that there were hardly any available datasets which contained the data needed for the training of the models. However, since a lot of research is being done in this area now, this has to the creation of custom-made data sets for this type of training models.

References

1. Cabani, A., Hammoudi, K., Benhabiles, H., Melkemi, M., MaskedFace-Net A dataset of correctly/incorrectly masked face images in the context of COVID-19. *Smart Health (Amst)*, 19, 100144, 2021.
2. Chiang., D., Detect faces and determine whether people are wearing mask, 2020. https://github.com/AIZOOTech/ FaceMaskDetection.
3. Loey, M., Manogaran, G., Taha, M.H.N., Khalifa, N.E.M., A hybrid deep transfer learning model with machine learning methods for face mask detection in the era of the COVID-19 pandemic. *Measurement (Lond)*, 167, 108288, 2021.
4. Khandelwal, P., Khandelwal, A., Agarwal, S., Using computer vision to enhance safety of workforce in manufacturing in a post COVID world, 2020. arXiv.org > cs > arXiv:2005.05287.
5. Wang, Z., Wang, G., Huang, B., Xiong, Z., Hong, Q., Wu, H., Yi, P., Jiang, K., Wang, N., Pei *et al.*, Y., Masked face recognition dataset and application, 2020. arXiv preprint arXiv:2003.09093.
6. Qin, B. and Li, D., Identifying face mask wearing condition using image super-resolution with classification network to prevent COVID-19, *Sensors*, 18, pp. 5236, 2020.

7. https://docs.microsoft.com/en-us/azure/machine-learning/component-reference/normalize-data
8. https://matplotlib.org/stable/index.html, Matplotlib 3.5.2 documentation.
9. https://machinelearningmastery.com/machine-learning-with-python/, Machine Learning Mastery with Python.

3

Application of Fuzzy Logic to Healthcare Industry

Biswajeet Sahu[1], Lokanath Sarangi[2], Abhinadita Ghosh[3] and Hemanta Kumar Palo[3]*

[1]Department of Electronics and Communication Engineering, YBN University Namkum, Ranchi, Jharkhand, India
[2]Department of Electronics and Communication Engineering, College of Engineering, Bhubaneswar, Odisha, India
[3]Department of Electronics and Communication Engineering, ITER, Siksha 'O' Anusandhan (Deemed to be University), Bhubaneswar, Odisha, India

Abstract

The healthcare industries need to deal with the uncertainty in the patient data, random reports, vagueness in medicines and prescriptions, a patient's mental and physical condition during testing, etc. The Fuzzy Logic (FL) can resolve these issues as it works similar to human minds based on approximate data. The algorithm removes redundancy from a massive database, summarizes, and develops crisp solutions. Its application in the healthcare decision-making process is fast developing due to the difficulties involved in designing and the development of a precise biological system model. The FL enforces researchers, medicos, and technology producers to explore smooth and creative solutions. It can be used as a versatile tool to model, manage, and control as our knowledge, and experience cannot be defined using explicit mathematical models. This motivates the authors to investigate the basic FL framework and analyze its application domains in this field. A descriptive study has been conducted to examine and focus on the healthcare decision-making process based on ranking, classification, data mining, feature selection, pattern recognition and optimization. It has also been addressed how FL-based strategies can be utilized to decrease uncertainty when working with simple and discriminative algorithms.

***Keywords*:** Fuzzy logic, healthcare, crisp solution, decision making, fuzzifier, inference engine

**Corresponding author*: hemantapalo@soa.ac.in

Sachi Nandan Mohanty, Prasenjit Chatterjee and Bui Thanh Hung (eds.) Fuzzy Computing in Data Science: Applications and Challenges, (37–54) © 2023 Scrivener Publishing LLC

3.1 Introduction

There are several context-depending scenarios where hard computing finds it difficult to provide an adequate model that can cater to all the requirements. For example, in a real-world environment such as healthcare, it is quite ambiguous to provide precise definition or meaning, or descriptions of many medical terms or relationships among important concepts accurately. The terms such as old, severe, long-time, certain point, etc. make the boundaries vague, unclear, and fuzzy. Similarly, many terms such as large, very large, extremely large, small, medium, very small, etc are relative and it is impossible to express them numerically [1, 2]. These factors make fuzzy logic an attractive alternative in the field of healthcare. Further, imperfect knowledge and the unpredictable nature of the available data lead to uncertainties in medical decision-making processes [3]. Some of the issues that make the development of effective healthcare identification system models can be viewed as below.

- Difficulty in understanding our biological mechanisms completely.
- The non-availability or ever-developing nature of test equipment.
- Difficulty in measuring the patient readings with precision.
- Ambiguous ranges of the standard value and test results.
- Presence of several conditions simultaneously.
- Missing of desired information and the presence of redundant data in many cases.

These factors make the traditional quantitative mechanism inadequate and complex. There always exists an inevitable and substantial amount of fuzziness while representing the biological system behaviour and its characteristics. This is due to the unavailability of an inappropriate mathematical framework to handle these entities involving a huge number of interrelated elements. Further, the presence of several variables makes these systems vague and fuzzy. Fuzzy sets can initiate notions of continuity into deductive thinking. In this way, these sets explore conventional symbolic systems in continuous form based on tabulated rules. As medicine exists in a continuous domain, the concept of continuous subset features such as fuzzy scores, fuzzy alarms, conventional scoring systems, etc. can accurately represent them. Similarly, fuzzy control is one such developed approach that utilizes a rule base to map the input into the output domain in continuous form. These logics can reason and quantify several ambiguous linguistic expressions in an appropriate form to alleviate imprecision and uncertainty.

In medicine, to address issues such as incompleteness, inconsistency, uncertainty, etc. it is not always essential to tackle micro-objectives and micro-phenomena. Due to a little bit of knowledge and it's inexplicit as well as contrasting behavior much is happening in medicine than in physical sciences [4]. This kind of problem may add value to healthcare industries. These can assist strategic decision-making occupying complex consequences requiring immediate medical attention. This kind of inaccuracy or incompleteness can be envisaged in several ways such as:

- Information and knowledge about patients with varying categories, and attributes which is bound to be unpredictable.
- A patient's case study and medical history described by him or her which may be highly subjective, exaggerated, simulated, or understated. Further, the failure to mention past operations, ignorance, or recollections of earlier symptoms or diseases often creates confusion or dents in successful diagnosis or treatment.
- Physical observation, in this physician, collect the data. They are doing errors and neglecting precious instructions, or also forgot to observe the whole test. There are some limitations between the regular and laboratory statement.
- Sometimes measurement errors in pathological tests, imperfect calculations, failure to focus on vital indications, overlooking minute details, and incomplete tests often mislead the medical practitioner. Similarly, faulty equipment, poorly managed and developed labs, inadequate knowledge, failure to interpret the range of standard and pathological status aggravate uncertainty.
- The accuracy of the ultrasonic, x-ray, and others laboratory examinations is highly dependent on the efficiency and knowledge of the lab technicians, reliable interpretation by the medical examiner, the behavior of the affected person before the samples are taken, optimal functioning of the testing equipment, etc.

Some of the most essential information required to evaluate a patient's health situation lies in how accurately an examiner interprets the laboratory test results, medical history, and a patient's behavior. It is highly unpredictable to conclude during an initial examination either physically or analytically, thus warranting further study on patients' condition. There are many variations in reports, categorization of testing techniques, interpreted data, etc. makes the diagnosis vague and leads to uncertainty. Thus, the application of FL techniques has been widely explored in anaesthesia, internal medicine, electrophysiology, radiology, pharmacokinetics, etc. to model and control data. Further, it can help in diagnosis and prediction of disease based on data on asthma, cardiac heart, Alzheimer's, diabetes, etc. using remote monitoring systems to help patients in distress. The sensor data on a patient's health condition during standing, walking, jogging or running, sleeping or laying, and other activities can provide vital inputs to develop effective models using FL. Such analysis can accommodate patient care with comfort, safety, and convenience by the use of labelled data on a numerical scale. The objective of the fuzzy models is to develop effective recognition health monitoring models at lower cost and precision. FL decision-making healthcare models can determine the type and level of affected disease using data mining algorithms to patients irrespective of their age or gender. The latest updates of the patient's condition, its severity, and the type of treatment to be prescribed based on the age can be estimated using the Fuzzy IF-THEN rules. There have been many soft computing approaches such as the Artificial Neural Network (ANN), Bayesian Network (BN), Fuzzy Inference System, Swarm Intelligence, Genetic Algorithms (GAs), and Fuzzy Cognitive Maps (FCMs) explored effectively to remove uncertainty and vagueness in this field [1]. Further, the application of FL can be extended to many service sectors to categorize analytic Hierarchy Processes that are reliable, tangible, responsive, and can assure empathy among medicos. Finally, the combination of Engineering and FL-based techniques can be applied to identify risks due to cancer, heart failure, kidney malfunction, diabetics, birth abnormalities, surgical complicacy, etc. in near future.

This piece of work analyses the application of Fuzzy Logic (FL) in healthcare industries based on earlier researches carried out in this field. The objective is to investigate and develop a concrete roadmap to forecast further developments in this field using the concept of fuzzy technology. Such analysis can provide vital clues on the application of FL techniques in healthcare industries to assist the decision-making process and remove uncertainty [5, 6]. Its capability to define relevant relationships set in a data system can assist the diagnosis, prediction, and treatment of the findings in several clinical outcomes [3]. The technology can be effectively applied to many vivid medical application domains which may be invasive, conservative, regional, neurological, basic sciences, signal and image processing, oriental, laboratory, nursing, etc. The focusing area mostly lies on the ability of the FL technology in handling uncertainty and improving decision-making in the field of medical industries.

3.2 Background

During the last few decades, the healthcare industries have grown leap and bound as compared to other computer-aided technologies. The exponential rise in medical industries and its vagueness has warranted the application of advanced technologies and creative strategies including FL modelling and artificial intelligence. The demand for FL techniques rises due to its extended application of the classical Boolean algebra in computer-aided healthcare applications. The technique connects the concepts to symbols, deals with the semantics of a corresponding domain, compares the associated constraints, and extends the particularizations similar to human reasoning [4]. The theories associated with Fuzzy set and fuzzy logic help to characterize uncertainty often encountered in the formation of medical concepts, interpretation of a patient condition, diagnosis of complex disease, and making therapeutic decisions. These Fuzzy sets are capable of defining the medical symptoms, test results, signs, diseases, diagnoses, prognostic, and, therapeutic information at ease and accurately, by conserving the obvious vagueness. Further, these logics can draw strict and approximate inferences reasonably. The technics suits a broad range of healthcare domain as mostly medical concepts requires causal, definitional, and heuristic, and statistical, knowledge. Sometimes, practical medicine has to accommodate incomplete, uncertain, vague empirical medical concepts to decide several medical procedures. Similarly, fuzzy automata can be utilized in real-time to deal with patient monitoring devices by accessing real-time medical information systems. It can be inferred that the use of FL and Fuzzy set theory can be further extended to develop knowledge-based healthcare systems such as interpreting the medical findings, disease diagnosis, prescribing optimal medical treatments, and monitoring the real-time patient data. A few applications of FL technologies have been explained below.

In Vienna General Hospital ICUs, the FuzzyKBWean has been used to control and optimize the ventilation and weaning process of patients during cardiac surgery. This is an open-loop fuzzy control system that accessed the Patient computer data management system fixed near the patient's bed [5]. The FuzzyKBWean rules consider the patient's physiological parameters as well as the actual ventilator settings using fuzzy antecedents and propose an appropriate new ventilator setting in time based on the rules. Correspondingly, the cooperating physicians can directly implement knowledge base changes in the ICU using the FuzzyKBWEdit rule editor [6].

To monitor patients' data in the ICU intelligently, the FuzzyARDS/MONITOR has been used successfully [7]. It detects the acute respiratory distress syndrome (ARDS) of patients timely and accurately to provide adequate therapeutical advice. The fuzzy automata remain a suitable approach in this case since the ARDS is a medical entity very ill-defined. In this technique, the ARDS states consider as pathophysiological entries corresponding to several states of applicable therapies. Thus, the affected individuals can be assigned to one or more domains partially. The fuzzy states are assumed to be false or true or partially true while transiting from one state to another. These fuzzy states are used to represent several high-level medical entities which may include trends of linguistical expressions. These can be evaluated permanently using the steps of data-to-symbol conversion with an adjustable time granularity. Similarly, the FuzzyARDS/STUDY is a web-based study system that facilitates the entry of data at the different study centres which may include the fuzzy criteria, fuzzy scores, and evaluation of patient data using interval techniques to determine missing variables in the available repository.

In this regard, The CADIAG project helps several diagnostic processes of internal medicine using computer-aided technologies. It deals with many clinical issues such as identifying different diseases based on the pathological findings, further prescribing essential examinations to gain extended diagnostic hypotheses and to validate the supported findings, and searching for unaccounted patient's pathological findings [8]. The CADIAG in Vienna General Hospital is fully integrated into its WAMIS medical information system. Its successor is named as MedFrame/CADIAG-IV which offers the positive and negative diagnostic hypotheses as well as several therapeutical proposals. The core of this technique include an integrated system that include the patient information and knowledge base system, editor modules, differential therapeutical and diagnostic modules as well as modules of immediate case evaluation.

3.3 Fuzzy Logic

The basic components of FL have been shown in Figure 3.1.

Fuzzy Logic has been applied in several home automation systems such as razors, Elevators, medical devices, subways, risk analysis in bank credit

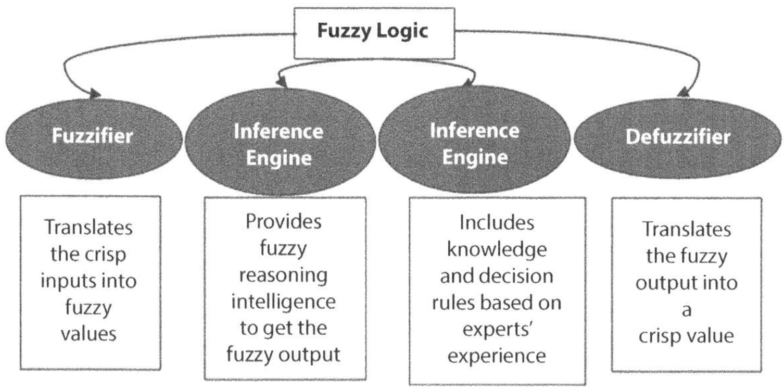

Figure 3.1 The basic components of FL.

transactions, bond ratings, etc. It considers the approximate rather than the crisp values or values under certainty. In the input space, the logic assigns a number or a membership function to every element . The membership function (MF) aims to map each input value to a corresponding membership value which indicates its membership degree in a fuzzy set. The MFs assign the values in an interval of 0 and 1 [9]. Among several shapes of MFs, linear, trapezoidal, triangular, and Gaussian are most popular. However, the triangular MF is frequently chosen due to its ease of use and practicality [10].

The fuzzifier in FL does the fuzzification a process that converts a crisp entity in a fuzzy set to a grade of MFs representing the linguistic variables [10, 11]. The obtained fuzzy sets on fuzzification are then processed by the Inference Engine (IE) a processing unit based on the rule base. The IE is an important module of the FL system and the knowledge processor which draws conclusion or inferences based on the existing data and the expert's reasoning. It maps a chosen input to the desired output using fuzzy sets. The knowledge base comprises the rule base and the database to determine the performance of an FL system [11]. The knowledge base contains several rules and uses the combination theory to choose the valid rules for the domain experts. A rule is considered to fire provided any of the parameters such as mild, moderate, severe, etc. is true else it does not fire.

The defuzzifier is used to translate the IE output into the crisp output. While the defuzzification input remains fuzzy the defuzzification output is crisp (a single number). The reason is usually the IE output is a fuzzy set, whereas while crisp values are mostly desired in a real-life scenario. The commonly used defuzzification mechanisms are

- max criterion
- centre-of-gravity
- the mean of maxima.

Figure 3.2 provides the advantages of the FL system while the disadvantages are provided in Figure 3.3.

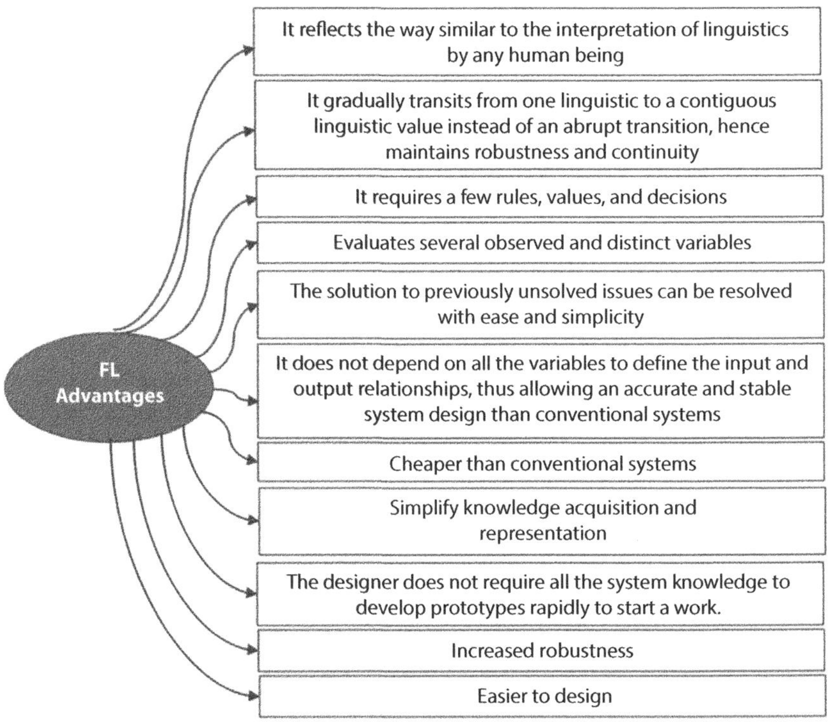

Figure 3.2 The advantages of a Fuzzy Logic system.

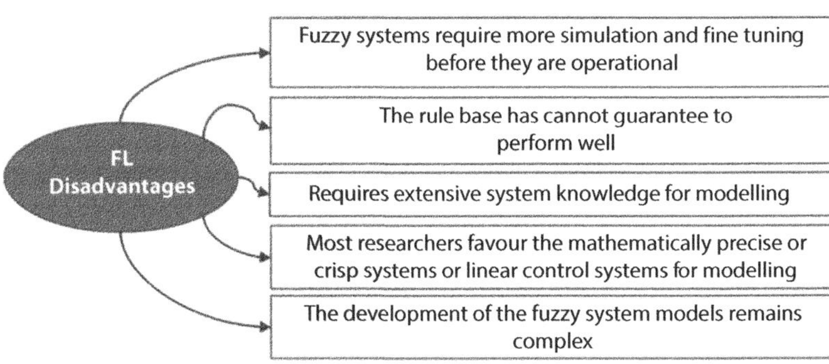

Figure 3.3 The disadvantages of Fuzzy Logic system.

3.4 Fuzzy Logic in Healthcare

The FL has been found applied in the field of healthcare either in single mode or hybrid mode. The hybrid mode is known as the neuro-fuzzy system which combines NN and FL or fuzzy plus Bayesian algorithm. The various ways in which the application of FL has been made are shown in Figure 3.4.

Clustering splits the data into clusters or groups and is a technique of data mining. It distinguishes the desired patterns by clustering them wherein similar features are kept in its corresponding cluster [10]. FL using the clustering techniques found its application in the field of images, cells, cancers, genes, etc. [11–13]. The Ranking is another FL mechanism to rank the risk factors, tests, suppliers, medicines, performance which assist in several strategic decision-making processes [14]. FL uses the classification mechanism group memberships for the predefined and known classes. This data mining FL technique has been found in many healthcare domains by several researchers [15–17]. In FL, the feature selection techniques aim to remove the redundant data and provide the relevant samples. This way, it reduces the memory space, increases the system response time, and releases the burden of the classifiers by preventing excessive training. The technique has been found to spot affected patients to get rid of unhealthy images, tissues, or cells [18, 19]. Pattern recognition is another FL application that deciphers the desired patterns in the corresponding subject. The authors used the time series analysis to determine the desired patterns based on time-dependent rules to identify medical images [20, 21]. In general, the medical data in the repository remains crisp. To extract vague information from the repository containing crisp requires advanced techniques such as Fuzzy query building, employing fuzzy ontology, Constructing fuzzy relation, etc. In Fuzzy query building, it is possible to

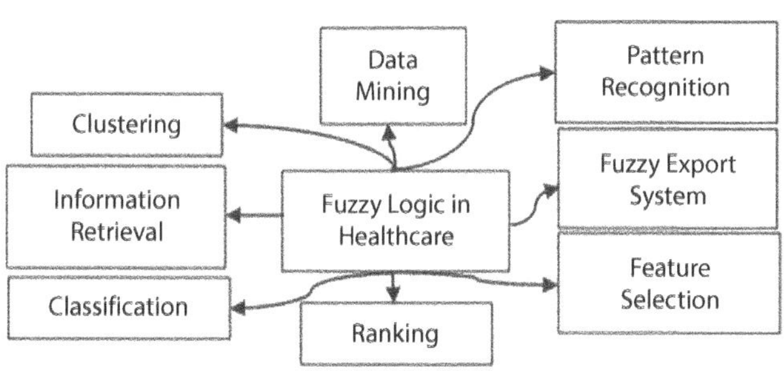

Figure 3.4 Fuzzy Logic in healthcare.

fulfil the individual information needs, crisp relational query languages' have been used though it remains inadequate [22]. For example, to develop a query like "exhibit the values of low liver enzyme for a child affected with celiac disease" we need to define the linguistic concepts in the language such as child or low. The objective of developing the queries to deal with complexity and impreciseness. Fuzzy SQL is an ORACLE-based fuzzy query language used to write flexible queries with the help of linguistic labels and fuzzy attributes in a crisp table.

Similarly, ontology can be used to represent a formal vocabulary specifying the domain knowledge as hierarchical concepts of structure, relations, and their attributes which can be read and understood by both man and machines. Thus, the medical ontology can conceptually represent the medical attributes in a symbolic way [23]. Nevertheless, these are based on the crisp concept, hence lack the desired elasticity or flexibility. It requires the application of FL in which the relationship among entities can be drawn based on the membership degree. The addendum of Database Management System (DBM) in FL is called fuzzy Relational DBMS in which the data in the database is represented using linguistic expressions. For example, the diabetics level in linguistic form can be represented as very low, or low, or high, or very high, etc.

The FL techniques have been widely explored in medical data mining applications. It uses data mining techniques to fetch both knowledge and information from the healthcare data repository for improving the quality and reducing the costs [24]. It is capable to deal with heterogeneous data involving symbolic, numerical, precise, approximate, ambiguous, and complexity. The FL techniques have been quite successful to describe these types of data with robustness being adaptable to the user's environment and variations in different parameters [24]. Several data mining mechanisms such as machine learning, decision trees, comparison measures, clustering, association rules, etc. include the FL concepts [25]. The application of FL for the diagnosis of diseases using the data mining techniques such as the ANNs, DTs, and Bayesian network has been attempted by earlier researchers [26]. The algorithm has been used for the diagnosis of cancer and heart diseases in Bayesian and ANN platforms [25].

The FL Cognitive Maps (CMs) must support decision-making in complex healthcare industries The system must fetch the desired causal knowledge and information from the medical data repository, develop the knowledge base, and must make suitable inferences [2]. The CMs are the knowledge description schemes that link the symbolic ontology approaches to the raw patient healthcare data [23]. They provide the relationships between cause and effect with ease and interpretable forms. The Fuzzy CMs are the dynamic Fuzzy models to describe the relationships among different medical concepts [23]. These Fuzzy CMs are dynamic, tunable, and combinable to experience the relationships among the medical concepts which are often vague, hidden, imprecise, and uncertain [27]. However, the application of Fuzzy CMs in healthcare industries requires defining the degree of influence of one concept on the other and vice versa. It requires historical data to observe the influences and their level of weights. The FL techniques have been explored often in internal medicine such as gastroenterology, rheumatology, pulmonology, and hepatology [27, 28]. In these studies, Fuzzy labels are derived using soft matching techniques and are used to provide the relationships among the diagnoses and signs of human health status. It has been applied in expert systems such as the CADIAG in which the medical data available in the information system has been used to provide the desired medical information. This information is then fuzzified based on the fuzzy rule base. The advanced CADIAG-2 has been found applied in the medical domains such as gastroenterology, rheumatology, and hepatology [28]. There are other FL-based expert systems found to be successfully explored are the RENOIR, SPHINX, and CLINAID. Figure 3.5 lists some of the application of FL in healthcare industries.

There have been considerable researches carried out to apply the FL in the evaluation and investigation of cardiovascular diseases, cardiac functions, cerebrovascular disease, ECG, cardiovascular medicine, artificial heart control, etc. [29–31]. TOTOMES is another FL application area designed and developed to determine the dynamics of cardiovascular disease during ventricular assistance. The technique interprets in multiple platforms to identify and detects different organ systems dynamically

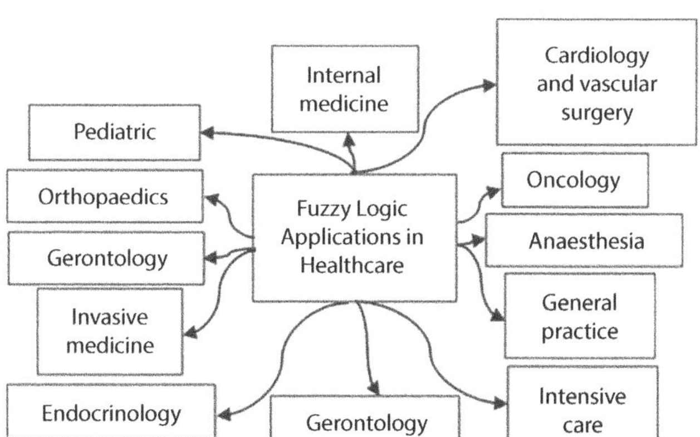

Figure 3.5 Lists of FL applications in healthcare industries.

using fuzzy reasoning to estimate a patient's health state. It can diagnose and detect the malfunctioning of a particular organ based on the fuzzy rule base. For example, tomography images have been associated with the FL techniques to classify the patients' heart function with 94% accuracy [32]. The analysis and investigation of coronary artery disease, ECG classification, cardiac disease identification are other several applications discussed in the literature [28]. In pediatric ICU, the FL expert system has been used as a warning mechanism by regularly assessing the EEG abnormality level [33]. These automated systems can distinguish the critical condition of the patients and their artifacts. The burdens on the diagnosis of cardiopulmonary emergencies in ICUs can be tackled using improved patient monitoring systems in real time. One such FL expert system is the FLORIDA which uses the knowledge bases to determine a patient's physiological condition in ICU. Similarly, the systems can be further extended to other healthcare domains such as breath detection and non-invasive sensor fusion [34]. It has been utilized to investigate the interpolation and the patterns of unborn babies based on the cardiogram reports which can prevent fatal injury or unnecessary medical complications [35]. Although, little work can be found on the application of FL techniques in Endocrinology or the determination of thyroid, the knowledge-based system has been widely used to monitor diabetics with hierarchical neural networks [36]. In oncology, the FL is concerned to classify discriminating cancer tissues, or breast tumours from the normal tissue with the help of 3-D ultrasonic echographic images for therapeutic advice [37].

The FL has been used to cluster Gerontology and in the veterinary expert system using several approaches corresponding to different subjects in a flexible way. The FL systems have been applied in determining the orthopedic symptoms for gait event identification of paraplegic subjects using electrically stimulated walking. There have been many reported pieces of literature in the application of FL techniques in anaesthesia to control the drug infusion, maintain adequate levels of anaesthesia, relaxation of muscles, patient alarm, and monitoring using virtual reality simulation and blood pressure [38]. The vague linguistic terms such as too long, or too close or too short, or too far can be used to measure the performance of the virtual reality surgical simulator. In this regard, the real-time expert system for decision making and control has been proposed to indicate the concentration of inhaled volatile anaesthetics [39]. The system has combined both the clinical data and online measurements in a Bayesian and fuzzy platform. Anaesthetists were confident enough to follow the dosage advice given by RESAC in most of the patients.

An application of this technology in plastic surgery to manipulate tools using an unsupervised segmentation method to create a new image of the patient has been another area worth noting in this direction [39]. Similarly, the technique helps to adapt and control the pharmacodynamic and pharmacokinetic parameter changes of the patient to prescribe treatments for muscle relaxation.

3.5 Conclusions

In this chapter, the application of FL in several fields of healthcare has been investigated. The natural disagreement among the medical staff, vagueness in medical concepts, uncertainty, and imprecise medical terms can be generalized using the fuzzy concept as revealed from our investigations. The techniques assist the computer and medical personal to learn and develop the desired concept with elastic modeling methods to accommodate themselves with smooth borders rather than approaching the decision models with binary approaches. As healthcare decision-making becomes more complex and tough, the exploration of FL techniques introduces creativeness to deal with the increased pace of developments. Different variations in the FL applications, such as ranking, clustering, data mining, classification, pattern identification, feature selection, and optimization have been discussed. The study reveals that the techniques can be applied in every strategic decision making concerning the healthcare domain. These areas involve database management, supply chain, diagnosis, health data mining, information retrieval, etc. Although Fuzzy nature makes the conventional methods in medical decision-making suffer from elasticity, however, the following advantages make it more promising.

- Flexibility: FL considers every possible value be it deterministic or blurred nature.
- Robustness: The models developed using the techniques robust as compared to the conventional methods of system modeling.
- Efficiency: It uses the available data adequately and represents them using crisp values and linguistic representation, hence remains efficient.

There are a few limitations that still haunt the FL technologies and require further research. For example, it requires considerable time and effort to design and develop Fuzzy systems. This results in a higher computation time to obtain the desired output models. However, the healthcare domain is rapidly progressing with the advent of new medical technologies, machines, medicines, equipment, and findings. There is an exponential rise is in supporting and allies' branches that contribute to this field enormously with medical informatics. The funds and donations to healthcare industries also increase each day to support decision-making processes that require further studies in this area. Thus, the field cannot deny the benefits of FL in solving healthcare issues on a large scale. The integration of FL with artificial intelligence, robotics, image, and speech processing are a few trending topics the future researchers need to consider. The advancement in new, expensive, sophisticated image and voice processing techniques provides better quality and minute details of images concerning EEG, ECG, MRI, CT scan, etc., which increases the applicability of FL. Medical imaging remains a vast industry, wherein a huge amount of money is being spent, which creates the space for FL. There is every possibility of an increase in research funding in solving uncertainty and vagueness using the Fuzzy techniques in the medical domain. The Fuzzy imaging techniques and FSQL with ORACLE compatible make future research lucrative and promising. Due to competence, fuzzy database management is going to further escalate in the future. Similarly, the hybrid techniques involving ANN, FL, machine learning, and artificial intelligence are going to play an important role in determining future decision making in healthcare sectors.

References

1. Bhatia, A., Mago, V., Singh, R., Use of soft computing techniques in medical decision making: A survey, in: *2014 IEEE International Conference on Advances in Computing, Communications and Informatics (ICACCI)*, pp. 1131–1137, 2014.
2. Gürsel, G., Healthcare, uncertainty, and fuzzy logic. *Digit. Med.*, *2*, 3, 101, 2006.
3. Yardimci, A., Soft computing in medicine. *Appl. Soft Comput.*, *9*, 3, 1029–1043, 2009.
4. Barro, S. and Marín, R., A call for a stronger role for fuzzy logic in medicine, in: *Fuzzy logic in medicine*, pp. 1–17, Physica, Heidelberg, 2002.
5. Schuh, C., Hiesmayr, M., Adlassnig, K.P., Zelenka, C., Klement, E.P., Integration of crisp-and fuzzy-controlled weaning in an ICU PDMS, in: *Proc. of the World Automation Congress*, WAC, vol. 98, pp. 299–304, 1998.
6. Koller, W., Schuh, C., Zelenka, C., Hiesmayr, M., Adlassnig, K.-P., *The fuzzy set and rule editor KBW edit of the medical expert System KBWean*, vol. *18*, pp. 27–29, OEGAI-Journal, Vienna, Austria, 1999.

7. Bernard, G.R. *et al.*, The American-European Consensus Conference on ARDS. Definitions, mechanisms, relevant outcomes, and clinical trial coordination. *AJRCCM*, *149*, 3, 818–824, 1994.
8. Leitich, H., Kiener, H.P., Kolarz, G., Schuh, C., Graninger, W., Adlassnig, K.P., A prospective evaluation of the medical consultation system CADIAG-II/RHEUMA in a rheumatological outpatient clinic, in: *Methods of information in medicine*, vol. *40*, pp. 213–220, 2001.
9. Zadeh, L. A., Fuzzy logic, neural networks and soft computing. In Safety evaluation based on identification approaches related to time-variant and nonlinear structures, Vieweg+ Teubner Verlag, 320-321, 1993.
10. Awotunde, J.B., Matiluko, O.E., Fatai, O.W., Medical diagnosis system using fuzzy logic. *Afr. J. Comput. ICT*, *7*, 2, 99–106, 2014.
11. Jantzen, J., *Tutorial on fuzzy logic, Technical Report, Dept. of Automation, Technical University of Denmark*, Pub. No. 98-E-868, 1998.
12. Ivezić, D., Tanasijević, M., Ignjatović, D., Fuzzy approach to dependability performance evaluation. *Qual. Reliab. Eng. Int.*, *24*, 7, 779–792, 2008.
13. Dweiri, F.T. and Kablan, M.M., Using fuzzy decision making for the evaluation of the project management internal efficiency. *Decis. Support Syst.*, *42*, 2, 712–726, 2006.
14. Dadone, P., *Design optimization of fuzzy logic systems*, Doctoral dissertation, *Virginia Polytechnic Institute and State University*, 2001.
15. Sharma, N., Bajpai, A., Litoriya, M.R., Comparison the various clustering algorithms of weka tools. *Facilities*, *4*, 7, 78–80, 2012.
16. Thong, N.T., HIFCF: An effective hybrid model between picture fuzzy clustering and intuitionistic fuzzy recommender systems for medical diagnosis. *Expert Syst. Appl.*, *42*, 7, 3682–3701, 2015.
17. Priya, D.K., Krithiga, S.R., Pavithra, P., Kumar, J.R., Detection of leukemia in blood microscopic images using fuzzy logic. *Int. J. Eng. Res. Sci. Technol.*, *240*, 197–205, 2015.
18. Wu, Y., Duan, H., Du, S., Multiple fuzzy c-means clustering algorithm in medical diagnosis. *Technol. Health Care*, *23*, s2, S519–S527, 2015.
19. Shrief, M., Al-Atabany, W., El-Wakad, M., Ranking and evaluating CT departments by fuzzy logic. *Int. J. Comput. Appl.*, *122*, 8, 8–15, 2015.
20. Nguyen, T., Khosravi, A., Creighton, D., Nahavandi, S., Medical data classification using interval type-2 fuzzy logic system and wavelets. *Appl. Soft Comput.*, *30*, 812–822, 2015.
21. Sridhar, B., Reddy, K.V.V.S., Prasad, A.M., Mammographic image analysis based on adaptive morphological fuzzy logic CAD system. *Int. J. Biomed. Eng. Technol.*, *17*, 4, 341–355, 2015.
22. Harikumar, R. and Kumar, P.S., Fuzzy techniques and aggregation operators in classification of epilepsy risk levels for diabetic patients using EEG signals and cerebral blood flow. *J. Biomater. Tissue Eng.*, *5*, 4, 316–322, 2015.
23. Kumar, S., Kashyap, M., Saluja, A., Bhattacharya, M., Segmentation of cotton bolls by efficient feature selection using conventional fuzzy C-means

algorithm with perception of color, in: *Proceedings of the Second International Conference on Computer and Communication Technologies*, Springer, New Delhi, pp. 731–741, 2016.

24. Alhaddad, M.J., Mohammed, A., Kamel, M., Hagras, H., A genetic interval type-2 fuzzy logic-based approach for generating interpretable linguistic models for the brain P300 phenomena recorded via brain–computer interfaces. *Soft Computing, 19*, 4, 1019–1035, 2015.
25. Banerjee, T., Keller, J.M., Popescu, M., Skubic, M., Recognizing complex instrumental activities of daily living using scene information and fuzzy logic. *Comput. Vis. Image Underst., 140*, 68–82, 2015.
26. Melin, P., Castillo, O., Kacprzyk, J., *Design of intelligent systems based on fuzzy logic, neural networks and nature-inspired optimization*, Springer International Publishing, pp. 593–603, 2015.
27. Tuben, U., Becks, A., Fathi, M., Tresp, C., Fuzzy queries in relational medical databases, in: *Proceedings of FUSION'98 International Conference*, FUSION, 1998.
28. Froelich, W. and Wakulicz-Deja, A., Mining temporal medical data using adaptive fuzzy cognitive maps, in: *2009 2nd Conference on Human System Interactions*, IEEE, pp. 16–23, 2009.
29. Mala, I., Akhtar, P., Zia, S.S., Mirza, S.H., Application of fuzzy relational databases in medical informatics, in: *2011 IEEE 14th International Multitopic Conference*, IEEE, pp. 41–44, 2011.
30. Bouchon-Meunier, B., Detyniecki, M., Lesot, M.J., Marsala, C., Rifqi, M., Real-world fuzzy logic applications in data mining and information retrieval, in: *Fuzzy logic*, pp. 219–247, Springer, Berlin, Heidelberg, 2007.
31. Delgado, M., Marín, N., Sánchez, D., Vila, M.A., Fuzzy association rules: General model and applications. *IEEE Trans. Fuzzy Syst., 11*, 2, 214–225, 2003.
32. Rotshtein, A. and Teodorescu, N.H., Design and tuning of fuzzy rule-based systems for medical diagnosis, in: *Fuzzy and neuro-fuzzy systems in medicine*, pp. 243–289, 1998.
33. Abbod, M.F., von Keyserlingk, D.G., Linkens, D.A., Mahfouf, M., Survey of utilisation of fuzzy technology in medicine and healthcare. *Fuzzy Sets Syst., 120*, 2, 331–349, 2001.
34. Kaufmann, R., Reul, H., Rau, G., The Helmholtz total artificial heart labtype. *Artif. Organs., 18*, 7, 537–542, 1994.
35. Joly, H., Sanchez, E., Gouvernet, J., Valty, J., Applications of fuzzy set theory to the evaluation of cardiac functio. *Proceedings of the MedInfo*, vol. *80*, pp. 91–95, 1980.
36. Kalmanson, D. and Stegall, F., Cardiovascular investigation and fuzzy concepts. *Am. J. Cardiol., 35*, 30–34, 1975.
37. Ota, D., Loftin, B., Saito, T., Lea, R., Keller, J., Virtual reality in surgical education. *Int. J. Biomed. Comput.*, 36, 4, 281–291, 1995.

38. Raposio, E. *et al.*, An "augmentedreality" aid for plastic and reconstructive surgeons, in: *Medicine meets virtual reality, global healthcare grid*, pp. 232–236, IOS Press, Amsterdam, The Netherlands, 1997.
39. Greenhow, S.G., Linkens, D.A., Asbury, A.J., Pilot study of an expert system adviser for controlling general anaesthesia. *Br. J. Anaesth.*, 71, 3, 359–365, 1993.

4

A Bibliometric Approach and Systematic Exploration of Global Research Activity on Fuzzy Logic in Scopus Database

Sugyanta Priyadarshini* **and Nisrutha Dulla**

KIIT Deemed to be University, Odisha, India

Abstract

Fuzzy logic stands as a viable approach toward models of reasoning and logic ranging from approximating human reasoning, delivering faster decision, manipulating uncertain statistics to paralleling the tradeoff between accuracy and significance. The objective of the current research work is to provide a structural overview and systematic review of the voluminous research work conducted on application of fuzzy logic in diverse dimensions and to assist the researchers in making insights into the subject. Bibliometric analysis is put into use to extract 493 records from Scopus database as a "CSV" file from 1990 till June 10, 2021, out of which 481 articles written in English language is taken into consideration. The extracted data are explored through Vos viewer to put light on eminent contributing countries and authors, frequent author keywords, cited references, total link strength, and co-occurrence of author keywords. Microsoft Power bi is used for graphical illustration of records. Out of 481 articles, 49.48% are research articles, 38.04% are conference papers, and 12.47% are review articles. A sharp peak in the year 2020 is discerned with a highest publication count of 43 documents till date. However documents published in 2018 got maximum number of citations (n_917). The prominent keywords in recent literature on fuzzy logic includes fuzzy VIKOR, industry 4.0, DEMATEL, and Internet of things. However, major research activities using fuzzy logic is concentrated in the subject area of computer science (n_58.21%). Maximum number of documents is published by China (n_20.58%) with a highest count of total citation of 902. North China Electric Power University has successfully published maximum of 11 documents with 195 citations. Fuzzy logic is comprehensively used in modern control system for

**Corresponding author*: sugyanta.priyadarshini@kiit.ac.in; ORCID: 0000-0001-7660-6162

Sachi Nandan Mohanty, Prasenjit Chatterjee and Bui Thanh Hung (eds.) Fuzzy Computing in Data Science: Applications and Challenges, (55–78) © 2023 Scrivener Publishing LLC

imitating the human behavior in decision making in a faster pace. The reviewed data extracted from the analysis will be useful for the academic researchers to find out the fundamental patterns and dynamics in the research direction and further promote fuzzy analysis in the areas that are yet to be incorporated.

***Keywords*:** Fuzzy logic, bibliometric analysis, scopus, content analysis, systematic review

4.1 Introduction

The advancement of internet and communication Technology and internet has called for an expansion of the concept of fuzzy logic and extension of its derivative models and theories while dealing with the complex decision making process [1]. The concept of fuzzy logic and fuzzy sets came into limelight in 1965 when Zadeh LA affiliated to the University of California at Berkeley initiated his research work on it. He analyzed that the notion of fuzzy analysis ranges from classical Boolean logic (true or false) to multivalued logic (true or false or unknown or others) [2]. While extending his research work on understanding of natural language by computer, he realized that every problem of the Universe cannot be limited to the absolute terms of 0 or 1 [3]. Every problem cannot be described in binary terms, if so, then much data lying in between of 0 and 1 will remain unexploited [4]. Thus, whether computing the activities of life based on degree of truth instead of translating everything into usual "true" or "false" can resolve the problem is a question worth pursuing. Under such a paradox, the emergence of fuzzy logic could made it possible for the computers to understand the natural language processing technologies by allocating the statement with the degree of truth ranging from 0 (null) and 1 (truth) by using real numbers [5]. As fuzzy logic stems from the mathematical study of multivalued logic that believes in quoting the sets with subjective definitions, such as "tall," "large," or "beautiful" alike judgments of a human to any problem or decision making process. Thus, fuzzy analysis can be simply defined as a heuristic approach in providing more opportunities to mimic the real-life occurrences, which are not limited to absolute truth or falsehood rather somewhere in between, such as partial truth. By the use of fuzzy logic in analyzing of the degree of truth, one can produce accurate results with inaccurate data. It was observed that in 1990s, only 7% of the research articles were dealing with fuzzy concepts but the area of research gained significance after 1991 by reaching 40% of published work by taking into account the extension of the fuzzy concept, such as fuzzy numbers, operations with fuzzy numbers, fuzzy intervals, and relationships between fuzzy quantities. Further, the number of citations associated with the works also escalated exponentially [6, 7]. Gradually, core research on fuzzy concept discovered that the fuzzy set theory can be put into use in vivid domains, including computer science, machine learning, engineering, artificial intelligence, automation systems, mathematics, energy fuels, environment science, telecommunication, social science, medicine, and others. Further, computing methods established on the basis of fuzzy logic can also be helpful in developing smart system for decision making, optimization and identification. With the passage of time, fuzzy has got attention from engineers of all sectors (including electrical, mechanical, civil, bio medical, chemical, geological, aerospace, mechatronics, and others), social (economics, psychology and management) and natural (chemistry, physics, biology, earth science) scientists. Numerous research articles have focussed on the use of fuzzy logic in several applications, such as transmission systems, facial pattern recognition, vacuum cleaners, weather forecasting system, project risk assessment, and stock trading. 1980s marked the beginning of theoretical analysis and practical illustrations of fuzzy rule systems grounded on the concept of fuzzy data set as well as fuzzy learning [8]. The field of fuzzy analysis has extended its tentacles in different studies most significantly in fuzzy decision systems, fuzzy algorithms and environment, fuzzy semantics and fuzzy linear programming and other parameters. The gravity of the use of fuzzy logic is determined by the growing size of multi-disciplinary research involving fuzzy analysis and its applications. This vast usage of fuzzy logic motivated the researcher to conduct a bibliometric analysis and predict its future evolution in vivid dimensions [9, 10]. In particular, this study focusses on the use of bibliometric analysis in determining the most influential author and their work in the field of fuzzy logic, their affiliations, country, and publications in specified journal by taking into account the fuzzy logic as a keyword in the documents. By the help of bibliometric analysis, the quality, performance and impact of the article published in the specified journal is accessed. Further, it also analyzes the sources of information from Scopus database via PRISMA (Preferred Reporting Items for Systematic reviews And Meta-Analyses) statement line of investigation [11, 12]. Additionally, VOS viewer software is put into use to portray data related to co-authorship, bibliographic coupling of sources, authors, publications and countries, and co-citations in a bibliometric mapping [13].

It is evident that there are voluminous research works on bibliometric analysis related to fuzzy analysis. However, some of the authors focussed on providing an overview of fuzzy research whereas few concentrated on providing a better understanding about fuzzy concepts and its applications, and others on scientific usage of fuzzy logic in vivid dimensions. The current research work mostly focussed on providing the intellectual readers, scientific researchers and academic scholars to gain a quick insight of quantitative analysis of fuzzy sets and fuzzy logic in existing literature.

4.2 Data Extraction and Interpretation

The bibliometric analysis followed three steps of data exploration and interpretation ranging from (i) data hunt and suitable strategy, (ii) data pooling and screening, and (iii) analysis and interpretation, which is represented in Figure 4.1 for clarity. In stage 1, a bibliometric analysis is performed by identifying 493 documents from Scopus database in CSV form from 1990 till 10th June 2021. Scopus database is preferred over PubMed, web of science and other scientific database as it takes into a vast source of 23,700 peer reviewed journals, which is comparatively larger than other database [14]. In stage 2, the retrieved records are analyzed by using an internationally widely used free bibliometric analysis software known as Vosviewer (Visualization of Similarities) [15]. The retrieved data analyzed via Vosviewer reveals the significantly contributing authors, countries, author keywords, cited references, total link strength, and co-occurrence of author keywords in Stage 3. Microsoft Power bi is used for geometrical

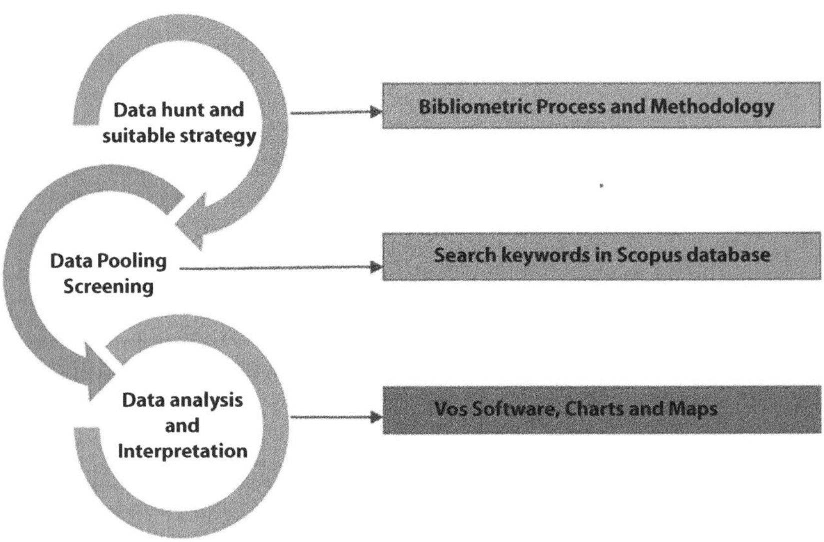

Figure 4.1 Stages of Bibliometric analysis for fuzzy logic.

representation of data. Search keywords are limited to articles, conference papers and conference reviews and further the language is restricted to English. However, 49.48% (n=238/493) are research articles, 38.04% (n=183/493) are conference papers, and 12.47% (n=60/493) are conference reviews. Out of 493 documents 481 are scripted in English language and 15.80% (n=76/493) are open access papers. Further, the assimilated records is applied in mapping the following: (1) per year publication and citation count, (2) prominent affiliations with maximum publications, (3) proactive journals covering wide spectrums, (4) prominent authors, (5) countries with major contributions, (6) co-occurrence of author keywords, (7) citation analysis of records, (8) cocitation analysis of cited references and cited authors.

4.3 Results and Discussion

4.3.1 Per Year Publication and Citation Count

Development in a particular area of research can be estimated by looking into the annual growth trend in publication count related it. At the same, the number of times the research work is followed by other sources can provide a brief about its quality. The higher the citations of a particular work, better is the work [16]. Figure 4.2 reveals the annual voluminous publications with citation count of research work using fuzzy logic from 1990 to June 2021 in a chronological order. However, it is observed that none of the years from 1990 to 2004 could come up with more than 10 publications yearly but from year 2005, more than 10 publications related to fuzzy logic were published.

The records retrieved from the database gained 5125 citations in total with an average cite score of 11 citations per document. The first two articles related to fuzzy logic was published in year 1990 but failed to gain any citation but the article published in the next year gained a total citation of 75. In year 1993 and 1997, there was not a single publication related to fuzzy logic. In the year 2018, highest number of articles related to fuzzy logic was produced (n=44) with a maximum count of citations (n=917). The article that gained maximum citations in year 2018 was "Awasthi A. (2018)" with 184 citations. The second highest publications was observed in year 2016 with 38 documents gaining a total citations of 479. The article that gained maximum citations in year 2016 was "Nilashi M. (2016)" with 71citations. Till June 10, 2021, 32 documents were published with a total citations of 23. Figure 4.2 represents the line graph drawn by using Microsoft Power bi

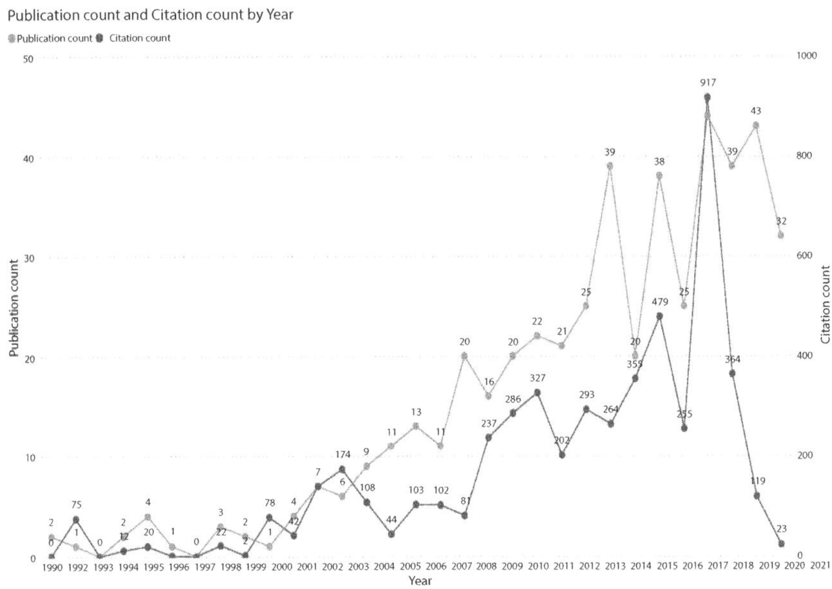

Figure 4.2 Yearwise publication and citation count of fuzzy logic.

that comprises two curves representing total publication count (red) and citations in total (blue). For overall count of documents, there is significant correlation between publication count in year and total citations of documents ($p<0.001$, $r=0.720$). Nevertheless, there is no significant correlation observed between the total publications and citations of the top sources ($p=0.088$, $r=-0.538$) and also there is no significant correlation observed between the total publications and citations of the top authors ($p=0.142$, $r=0.674$). This is evident that with the passage of time, the use of fuzzy logic is coming into limelight and is gaining citations, which depicts large interest in the research area.

4.3.2 Prominent Affiliations Contributing Toward Fuzzy Logic

North China Electric Power University produced maximum number of documents (n=11) with a total citation of 195 followed by two universities of Turkey, i.e., Galatasaray Üniversitesi with eight documents and 93 citation count and then by Istanbul Teknik Üniversitesi with six documents and 97 citation count. Two eminent authors naming Büyüközkan, G. (PC= 5, TC=85) and Göçer, F. (PC=4, TC=83) publishing highest number of publications are currently working in Galatasaray Üniversitesi. North China Electric Power University has started publishing research work on

Table 4.1 Prominent affiliations contributing toward fuzzy analysis.

Affiliations	Country	PC	TC	h-index	FPY	LPY
North China Electric Power University	China	11	195	8	2010	2021
Galatasaray Üniversitesi	Turkey	8	93	4	2006	2021
Istanbul Teknik Üniversitesi	Turkey	6	97	3	2010	2021
University of Tehran	Iran	6	119	4	2004	2020
National Cheng Kung University	Taiwan	5	124	4	2002	2010

Publication count (PC), Total citations (TC), First publication year (FPY), Last publication year (LPY).

fuzzy logic in 2010 and has already acquired maximum citations of 195 and Galatasaray University has started publishing research work related to fuzzy logic in 2006 and acquired 93 citations, Istanbul Teknik University has started publishing research related to fuzzy logic in 2010 and acquired 97 citations, University of Tehran has started publishing research related to fuzzy logic in 2004 and acquired 119 citations, National Cheng Kung University has started publishing research related to fuzzy logic in 2002 and acquired 124 citations. Table 4.1 puts light on the top 5 universities focusing on fuzzy logic on the basis of publication and citation count.

4.3.3 Top Journals Emerging in Fuzzy Logic in Major Subject Areas

In academics, more number of publications as well as research work on a specific topic signifies the necessity of it and the quality of the same represents better understanding on the subject matter encouraging thrust in the area of research work [17]. On one hand, the volume of research work can be predicted from the growing trend of publications in the particular area of research and on the other hand quality of research work in different journals can be judged from cite score (calculated on the basis of total citations acquired by the journal in the last 4 years divided by total publication count in the same time frame), total citations (number of times a particular idea is followed or preferred while writing a research work), Scimago journal rank (measures the prestige earned by the journal in a given year based on its weighted citations received by the articles published divided by its total publications in last three years), and source normalized impact per paper of the journal (evaluates the contextual citation impact via total number of citations in a specific theme) [18]. The 3 prominent journals that contributed toward eminent publications in the field of fuzzy logic are "Advances in Intelligent Systems and computing" with a count of 15 publications and 13 citations, followed by "Lecture Notes in Computer

Table 4.2 Top Journals publishing fuzzy-related works.

Sources	PC	TC	h-index	CS	SJR	SNIP	FPY	LPY
Advances In Intelligent Systems And Computing	15	13	2	0.9	0.184	0.428	2016	2021
Lecture Notes In Computer Science Including Subseries Lecture Notes In Artificial Intelligence And Lecture Notes In Bioinformatics	14	19	3	1.8	0.249	0.628	1995	2020
Communications In Computer And Information Science (DISCONTINUED)	12	0	0	-	-	-	2007	2020
Applied Mechanics And Materials (DISCONTINUED)	9	0	0	-	-	-	2013	2014
Expert Systems With Applications	8	463	7	12.7	1.368	3.079	2003	2020
IEEE International Conference On Fuzzy Systems	8	89	5	1.8	0.280	0.557	1995	2017
Advanced Materials Research (DISCONTINUED)	7	3	1	-	-	-	2012	2014
Soft Computing	6	119	4	5.1	0.626	1.463	2003	2021
Energy	5	156	5	11.5	1.961	2.014	2015	2021
Journal of cleaner production	5	89	3	13.1	1.937	2.475	2016	2021
Journal of intelligent and fuzzy systems	5	57	3	2.8	0.331	0.686	2012	2019

CS (Cite score), TC (Total citations) SJR (Scimago journal rank) SNIP (Source normalized impact per paper of the journal) PC (Publication Count), TC (Total Citations).

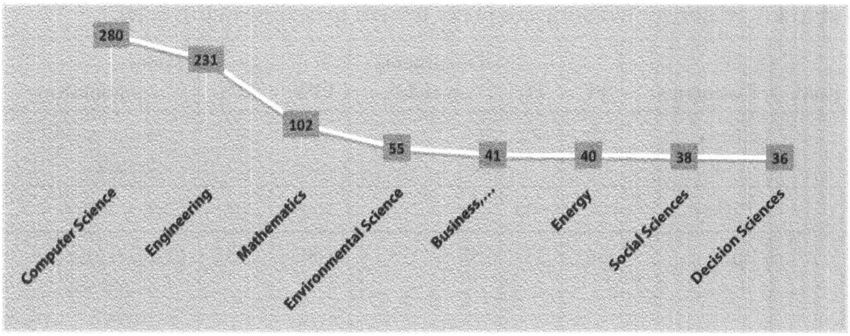

Figure 4.3 Major subject areas using fuzzy logic.

Science Including Subseries Lecture Notes in Artificial Intelligence and Lecture Notes in Bioinformatics", with a count of 14 publications and 19 citations and "Communications in Computer and Information Science" with a count of 12 publications but unfortunately no citations for it. Further, Table 4.2 portrays the major sources of publications with their Total citations, Journal cite score, Scimago journal ranking, h-index, and Source normalized impact per paper.

However, the 8 major subject areas using fuzzy analysis for publishing maximum articles are computer sciences with 34.02%, followed by Engineering with 28.06 %, mathematics with 12.39%, environmental science with 6.68% and business management with 4.98, Energy with 4.86%, Social science with 4.61% and Decision Science with 4.37 which is depicted in Figure 4.3.

4.3.4 Major Contributing Countries Toward Fuzzy Research Articles

On the basis of data collected through bibliometric analysis from Scopus database, a list of prominent countries contributing voluminously toward fuzzy logic is projected in Table 4.3.

China tops the list with maximum publication count (n=99) and citation count (n=902) followed by Turkey (PC=37, TC=633) India (PC=36, TC=331), Taiwan (PC=34, TC=647), United States (PC=29, TC=438), Iran (PC=22, TC=447), Canada (PC=19, TC=501), Australia (PC=16, TC=227), United Kingdom (PC=16, TC=172), Italy (PC=11, TC=263). It is observed that even if both Australia and United Kingdom have similar number of publications but Australia exceeds United Kingdom in Citation count by 55. Further, even if Taiwan ranks 4th in publication count but has

Table 4.3 Major contributing countries toward fuzzy research articles.

Rank	Countries	PC	TC	Documents h-index	FPY	LPY	Major collaborators
1	China	99	902	17	2001	2021	Canada, Australia, France
2	Turkey	37	633	14	2002	2021	China, Iran
3	India	36	331	10	2005	2021	China, United Kingdom, Turkey
4	Taiwan	34	647	13	2002	2021	France, China, United States
5	United States	29	438	10	1992	2021	China, Taiwan
6	Iran	22	447	10	2004	2021	Malaysia, Lithunia
7	Canada	19	501	11	2000	2019	Australia, Germany
8	Australia	16	227	6	2004	2021	China, Canada, Srilanka
8	United Kingdom	16	172	8	2001	2021	India, China, Hong Kong
9	Italy	11	263	5	1994	2021	Spain, United Arab Emirates

gained higher citations for its documents in comparison to Turkey and India. India ranks 3rd in publication count but its total citation is less than Canada (ranked 7th) which has a total of 19 publications. Table 4.3 puts light upon the international collaboration of these 9 prominent countries on fuzzy logic. It was observed that China had major collaborations with Canada, Australia and France. However, Australia has major collaboration with China but Canada has limited it with Australia and Germany. International collaboration with foreign countries enhances country wide team up to access greater knowledge, skills and expertise in vivid themes of research work, which has every potential to build a strong problem solving strategy to build a robust research base.

On the basis of the data collected from Scopus database, the co-authorship of countries, network visualization is created (Figure 4.4) via Vosviewer. In the process of mapping, minimum number of documents of a country and citations is limited to 2 each. Minimum number of

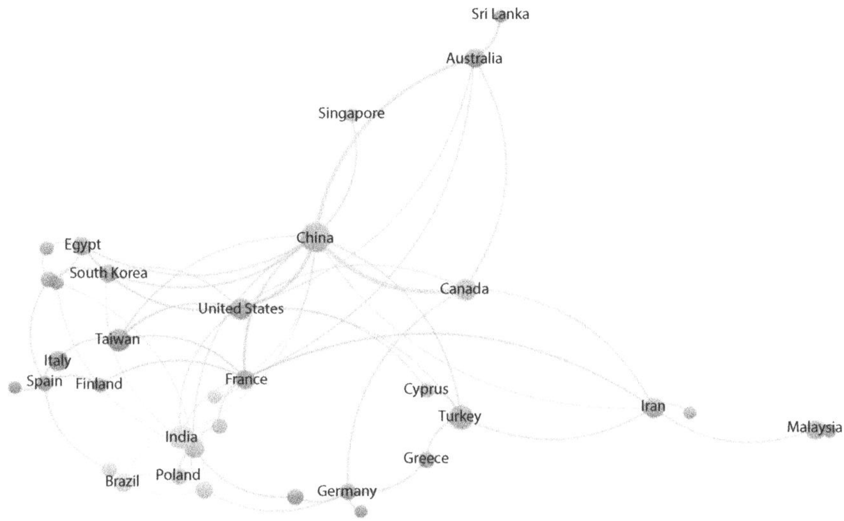

Figure 4.4 Overlay visualization of countries.

documents of a country limited to 2, minimum number of citations limited to 2, out of 73 countries 44 met the threshold. For each 44 countries the total strength of co-authorship link with other countries is calculated. The countries with greatest total link strength is then selected. Only 35 countries are connected to each other. 35 countries were segregated into eight clusters. Cluster 1 (comprising of six items) Colombia, Finland, France, Italy, Spain, Taiwan followed by cluster 2 (comprising of six items) Czech republic, Egypt, Pakistan, Saudi Arabia, South Korea, United states then cluster 3 (comprising of six items) Cyprus, Germany, Greece, Hong Kong, Romania, Turkey, and cluster 4 (comprising of five items) Brazil, India, Morocco, Portugal, Russian federation, then cluster 5 (comprising of four items) Bangladesh, Iran, Lithonia, Malaysia, then cluster 6 (comprising of three items) Canada, China, Singapore, and cluster 7 (comprising of three items) Mexico, Poland, United Kingdom and ultimately cluster 8 (comprising of two items) Australia and Sri Lanka.

The size of circle represented in Figure 4.3 shows the number of documents published by the particular country. Larger the size of the circle, more is the number of documents published by the country. The countries that are represented with same color implies that these countries have more co-authored documents. Through overlay visualization it was found that Russian federation, Morocco, Saudi Arabia had started publishing articles related to fuzzy logic with an average publication year 2018.

4.3.5 Prominent Authors Contribution Toward the Fuzzy Logic Analysis

After taking into consideration all the records retrieved from Scopus database, it was discovered that approximately 1178 authors have contributed toward fuzzy analysis in vivid dimensions with a mean of 4.35 to at least 5 authors per publication. Büyüközkan, G. from Galatasaray University in Turkey has published maximum number of research articles (n=5) with highest citation count (n=85), document h-index (n=3) and author h-index (n=43) by publishing his first paper in 2006 and his last paper was published in 2021. 5 authors have produced 4 articles on fuzzy analysis in their research work. Out of those 5 authors, it was observed in Table 4.4 that Göçer, F. from Galatasaray University in Turkey has highest number of citation (n=83) by publishing his first article in 2017 and last article in 2021. Two authors from Università degli Studi di Napoli Federico II in Italy naming Acampora, G. and Loia, V. have equal number of publications (n=4) and citations (n=74) and h-index of documents (n=3) but Loia, V. exceeds Acampora, G. in author's h-index count but surprisingly both of them had published their first article in 2009 and lastly they published

Table 4.4 Renowned authors contributing toward fuzzy analysis.

Author	PC	Current affiliation	Country	TC	Documents h-index	Author's h-index	FPY	LPY
Büyüközkan, G.	5	Galatasaray Üniversites	**Turkey**	85	3	43	**2006**	**2021**
Acampora, G.	4	Università degli Studi di Napoli Federico II	**Italy**	74	3	22	**2009**	**2011**
Barin, A.	4	Federal University of Santa	**Brazil**	9	2	5	**2009**	**2011**
Canha, L.N.	4	Universidade Federal de Santa Maria	**Brazil**	9	2	16	**2009**	**2011**
Göçer, F.	4	Galatasaray University	**Turkey**	83	3	8	**2017**	**2021**
Loia, V.	4	Università degli Studi di Salerno	**Italy**	74	3	37	**2009**	**2011**

article related to fuzzy in 2011. Similarly, two authors from Brazil naming Barin, A. and Canha, L.N. have equal number of publications (n=4) and citations (n=9) and h-index of documents (n=2) but Canha, L.N. exceeds Barin, A. in author's h-index count whereas both are similar in first publishing year (2009) and last publishing year (2011).

4.3.6 Coauthorship of Authors

Figure 4.5 represents network visualization of the co-authorship of authors contributing toward fuzzy analysis. In this process, the minimum number of documents of an author is limited to 2 and minimum number of citations is limited to 10. However, out of 1178 authors only 60 met the threshold. For each of the 60 authors the total strength of co-authorship link with other authors is calculated. The authors with greatest total link strength is selected. Out of these 60 authors, the largest set of connected items were 9, these 9 authors are visible through network visualization. These nine items were further divided into three clusters. Cluster 1 comprises four authors naming LI, J; Liu, P; Wang, F; Zhang, J. Second, cluster 2 comprises three authors naming Lin z, Yin J and, Zhang L. Third, cluster 3 comprises two items authors naming Zhou, D; Zhou, P. Bigger the circle, higher is the number of documents produced by the author, such as Wang F, with highest number of publications is represented with a bigger circle in comparison to others. Different shades of the circle in over lay visualization helps in discriminating authors on the basis of their publication count and average publication year. The deep blue represents average publication year as 2016, the green color represents average publication year as 2018, and red color average publication year as 2017.

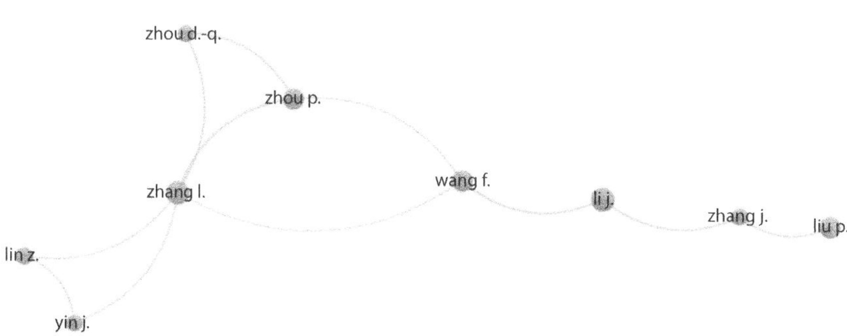

Figure 4.5 Network visualization of the co-authorship of authors contributing toward fuzzy analysis.

4.3.7 Cocitation Analysis of Cited Authors

The minimum number of citations of an author gained was fixed to 25 and as a result, out of 21,250 authors only 55 meet the threshold. For each of the 55 authors the total strength of co-citation links with other authors is calculated. The authors with greatest total link strength is selected. It was found that 54 items were largely connected. Further, those 54 items were divided into 5 clusters, with 1256 links and 999.75 as total link strength. Authors in similar cluster color in Figure 4.6 represents they have cited each other's works. It can be understood by documents in same cluster is linked to one specific area/one authors work is extension of others. Cluster red (23 items), green (12 items), blue (11 items), yellow (6 items), and purple (two items) represents top down order in terms of publication year and citations, i.e., cluster 5 in purple color with two authors Govindan K, and Smith S.F are new to the filed compared to other authors in other clusters.

4.3.8 Cooccurrence of Author Keywords

The bibliometric analysis conducted on fuzzy analysis identified 1314 keywords in the research work. The keyword fuzzy logic is recognized as the most used keyword with an occurrence of 54 times in the persisting literature followed by Decision making with 17 times, fuzzy Ahp by 15 times, Ahp by 12 times and decision support system by 11 times.

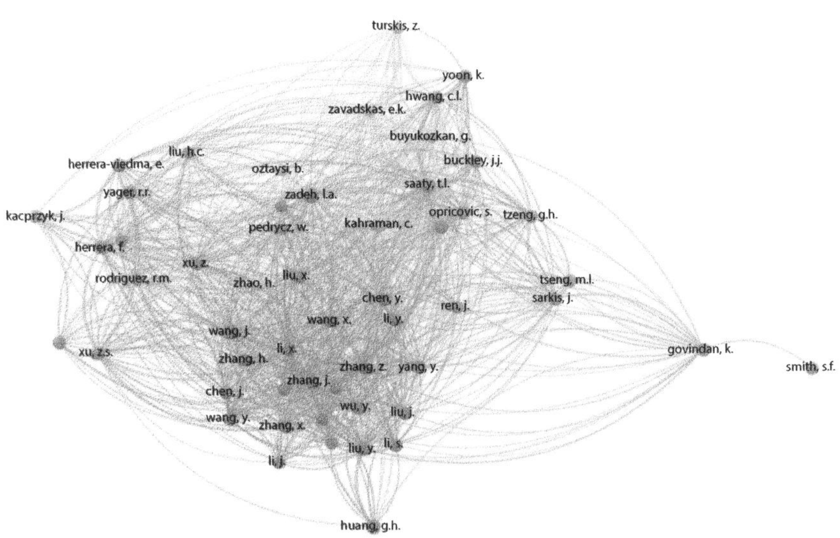

Figure 4.6 Cocitation analysis of cited authors.

To elucidate the research hotspots in fuzzy analysis, keywords cooccurrence is investigated via Vosviewer by using fractional counting. The co-occurrence threshold for the keywords is limited to minimum number of occurrence to five times. Out of 1314 keywords, only 32 meet the threshold. For each 32 keywords the total strength of co-occurrence link with other keywords is calculated. Those 32 keywords were divided into six clusters, having 106 links, and a total link strength of 84.50. It is evident that the new keywords in the latest research area are: dematel, industry 4.0, fuzzy vikor. These 32 keywords are illustrated in a network visualization in Figure 4.7. The relative size of circle in Figure 4.5 denotes the co-occurrences of the keywords. The larger the circle is, more the times a keyword is selected in the publications related to fuzzy analysis. The keywords fuzzy logic has total link strength of 25 which has linked o other 19 keywords. However, the strength and similarity between the themes is exhibited by the distance between the two keywords. Same colored circle shows the similarity of topic among diverse published articles. The co-keyword network visualization shown in the figure exhibits six

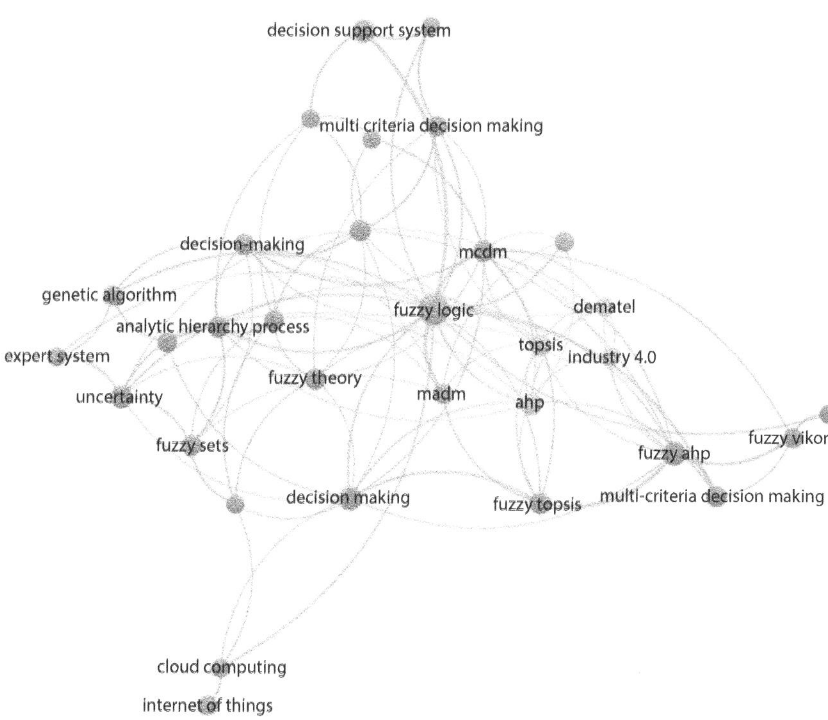

Figure 4.7 Network visualization of co-occurrence of author keywords.

different clusters. As a result, six main clusters are appropriately labeled on the basis of the node circles. Cluster 1 comprised 10 items (middle left in red color), such as analytic hierarchy process, decision making, expert systems, fuzzy, fuzzy sets, fuzzy theory, genetic algorithm, performance evaluation, supplier selection, and uncertainty. Next, cluster 2 comprised eight items (upper middle in green color), such as decision support system, decision making, fuzzy decision making, fuzzy inference system, group decision making, Mcdm, multi criteria decision making, and sustainability. Following, cluster 3 comprised five items (lower right in blue color), such as advanced manufacturing technology, fuzzy ahp, fuzzy topsis, fuzzy vikor, and multicriteria decision making. Further, cluster 4 comprised four items (middle right in yellow color), such as Ahp, dematel, industry 4.0, topsis. Additionally, cluster 5 comprised three items (center in purple), such as fuzzy logic, madm, network selection. Ultimately, cluster 6 comprised two items (lower left in sea blue), such as cloud computing and Internet of things.

4.4 Bibliographic Coupling of Documents, Sources, Authors, and Countries

4.4.1 Bibliographic Coupling of Documents

Out of 481 documents contributed by over all countries toward fuzzy analysis in Scopus database, a threshold limit of documents with a minimum of 10 citations is taken into consideration. 132 documents matched the threshold. For each 132 documents the total strength of bibliographic coupling links with other documents is calculated. The documents with greatest total link strength is selected. The largest set of connected items were 99. However, these 99 items were divided into 11 clusters. Cluster 1 (19 items in red color), cluster 2 (13 items in green color), cluster 3 (11 items in blue color), cluster 4 (nine items in yellow color), cluster 5 (9 items in purple color), cluster 6 (nine items in sea blue), cluster 7 (nine items in orange color), cluster 8 (eight items in brown color), cluster 9 (six items in pink color), cluster 10 (four items in light red), cluster 11 (two items in light green) is presented in Figure 4.8. The overlay visualization of the country wide documents is portrayed where the bigger circles represents comparatively higher citations than the ones with smaller circles. Highly cited documents are Colak M (2020) with 184 citations and total link strength of 13; followed by Wang X (2020) with 170 citations and total link strength of 4; Morente-Molenera J.A (2019) with 112 citations and total link strength of 3;

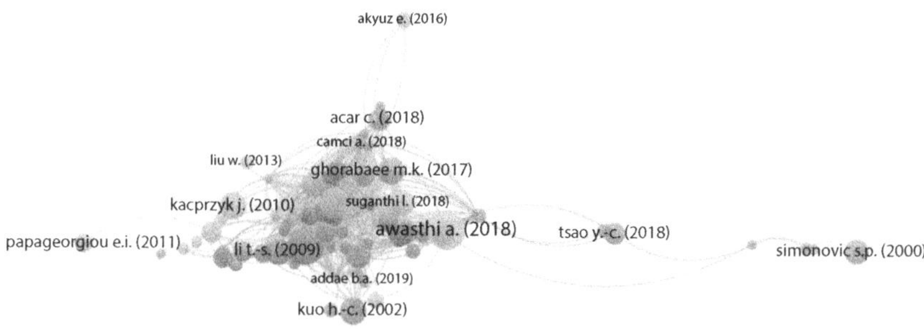

Figure 4.8 Overlay visualization of bibliographic coupling of documents.

and Addae, B.A (2019) with 110 citations and total link strength of 6 as shown in Figure 4.8.

4.4.2 Bibliographic Coupling of Sources

A total of 481 documents are published in 323 sources in Scopus database. A threshold limit of 2 documents with a minimum of 10 citations is taken into consideration. Out of 323 sources, 39 met the threshold. For each of the 39 sources, total 4l strength of bibliographic coupling with other sources is calculated. The sources with greatest total link

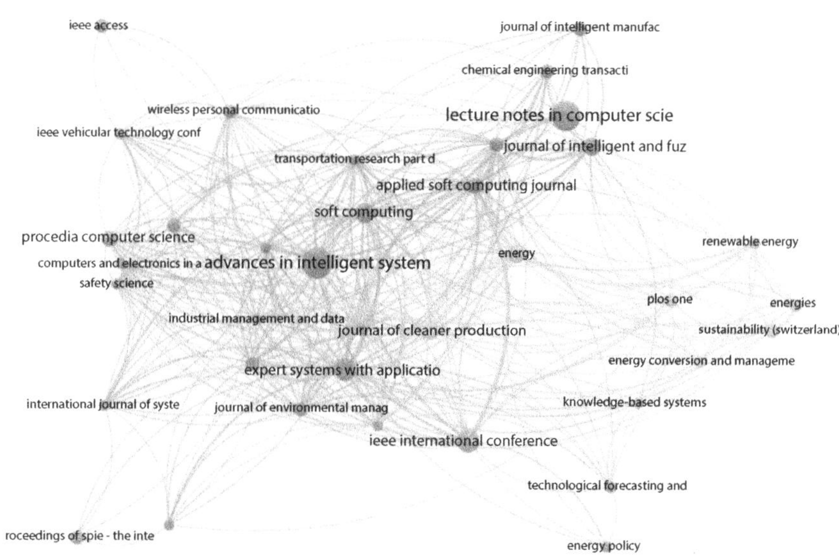

Figure 4.9 Overlay visualization of bibliographic coupling of sources.

strength is selected. The largest set of connected items were 37. These 37 items were divided into five clusters. Cluster 1 (10 items in red), cluster 2 (8 items in green), cluster 3 (7 items in blue), cluster 4 (7 items in yellow), cluster 5 (5 items in purple) is shown in Figure 4.9. Top sources are Brazilian power electronics conference Cobep 2009 with 1 documents, 463 citations and a total link strength of 27 followed by Chinacom 2010 with 1 documents, 184 citations and a total link strength of 3; then by with WCCI 2010 with 1 document, 156 citations and a total link strength of 7, further by with ICMCE 2010 with 1 document, 124 citations and a total link strength of 24, and ultimately wit heeeic.eu with 1 document, 119 citations and a total link strength of 8. However, Correlation is not significant between number of documents published in a source and its citation ($p=0.718$, $r=-0.023$ for $n=323$), but there is a significant relation between publication count and its total link strength ($p<0.001$, $r=0.412$ for $n=323$). Further, Correlation is also not significant between citation and its total link strength ($p=0.873$, $r=0.010$ for n=_323_).

4.4.3 Bibliographic Coupling of Authors

It was discovered that 1178 authors have contributed approximately 491 documents toward fuzzy analysis. A minimum number of documents for an author is fixed as 2 with a minimum number of citations as 10. Out of 1178 authors, only 60 meet the threshold. For each of the 60 authors, the total strength of bibliographic coupling links with other authors is calculated. The authors with greatest total link strength is selected. The largest set of connected items were 59. These 59 items were divided into 11 clusters. Cluster 1 (15 items in red), cluster 2 (9 items in green), cluster 3 (8 items in blue), cluster 4 (6 items in yellow), cluster 5 (4 items in purple), cluster 6 (three items in sea blue), cluster 7 (three items in orange), cluster 8 (three items in brown), cluster 9 (three items in pink), cluster 10 (three items in light red), cluster 11 (two items in light green) are presented in Figure 4.10. On one hand most productive authors with highest number of publications as well as highest number of citations are Acampora g with 5 documents and 152 citations (154 as total link strength), followed by Alahakoon D with four documents and 123 citations (189 as total link strength); Aviso K.B. with 4_documents and _113citations (86 as total link strength); Beskese A. with 4 documents and 108 citations (255 as total link strength). However, it is found that Correlation between total publications and citations count is significant ($p<0.001$, $r=0.880$ for $n= 60$), Correlation between total publications and total link strength is not significant ($p=0.080$, $r=0.228$ for $n=60$

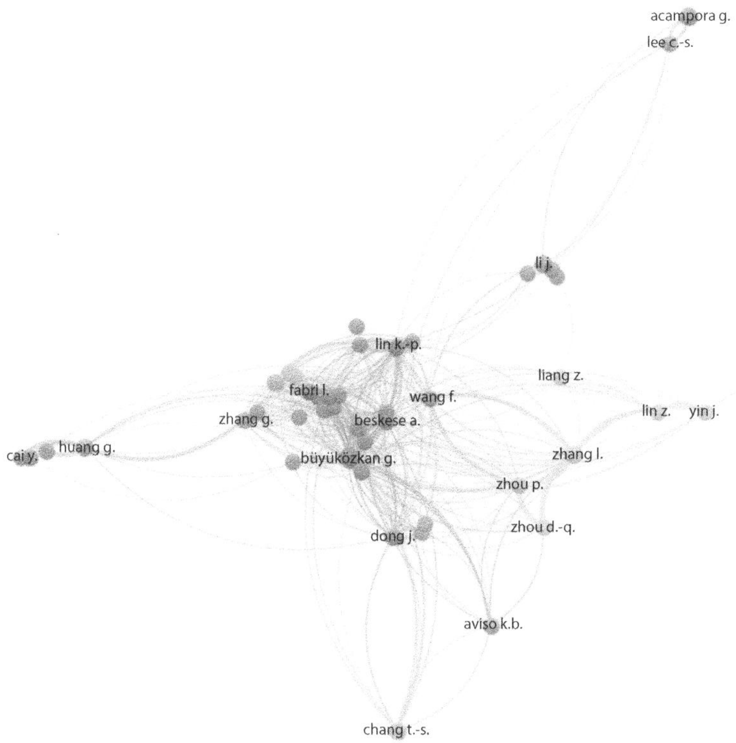

Figure 4.10 Overlay visualization of bibliographic coupling of authors.

and correlation between citation count and Total link strength is also not significant (p=0.569, r=0.075 for n=60).

4.4.4 Bibliographic Coupling of Countries

A total of 73 countries have contributed toward fuzzy analysis. A minimum number of documents per country is fixed as 2 with a minimum number of citations of 10. Out of 78 countries, 38 countries met the threshold. For each of the 38 countries, the total strength of bibliographic coupling links with other countries is calculated. The countries with greatest total link strength is selected. A total of 38 items were divided into 8 clusters links. Cluster 1 (seven items in red color), cluster 2 (seven items in green color), cluster 3 (six items in blue color), cluster 4 (five items in yellow), cluster 5 (five items in purple), cluster 6 (three items in sea blue), cluster 7 (three items in orange), cluster 8 (two items in light brown) is exhibited in Figure 4.11. Countries with Highly cited documents are Australia with

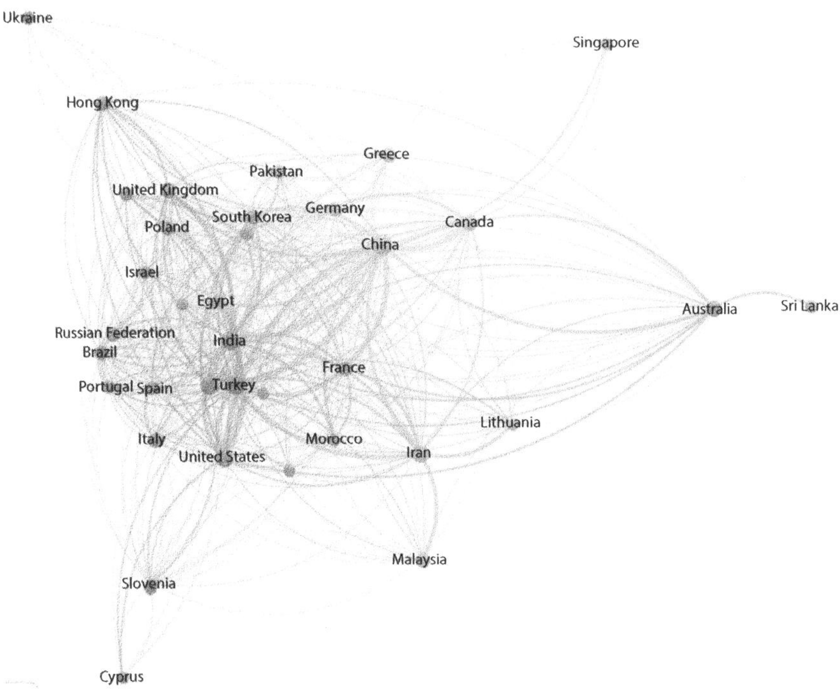

Figure 4.11 Overlay visualization of bibliographic coupling of countries.

902 citations and total link strength of 610 followed by Azerbaijan with 647 citations and total link strength of 5; Brazil with 10 documents, 633 citations and total link strength of 363; and Canada with 18 documents, 484 citations and total link strength of 885.

4.5 Conclusion

The research work has portrayed highly analyzed and statistically functional bibliometric analysis of fuzzy logic by extracting data from Scopus database regarding author's productivity, citations gained per paper, prominent journals, highly productive institutions and countries. VOS viewer is used to graphically illustrate co-citation analysis, co-authorship and bibliographic coupling of Documents, Sources, Authors and countries. Keeping an account of 495 publications in Scopus database on fuzzy analysis, conclusion that can be summarized are as follow:

1. There is a growing number of international collaboration and connection between authors, institutions and countries

working on fuzzy sets and fuzzy analysis in diverse dimensions. This inspires researchers of world concerned with fuzzy logic to come forward and work together to find out better results.

(ii) The graphical illustration of the data obtained from Scopus database and represented via VOS viewer made the network nodes between documents, authors, institutions, sources, and countries more prominent.

(iii) It was observed that fuzzy family comprises several concepts, such as fuzzy decision making, fuzzy inference system, fuzzy Ahp, fuzzy topsis, fuzzy vikor, and fuzzy logic. However, with the incorporation of fuzzy logic in decision making, a new area of research was attracted the researchers of the world toward it. Thus, the concept of fuzzy logic accounts for approximately 40% of documents of total documents related to fuzzy analysis till 2020. This projects the interest of the researchers in the topic of fuzzy logic. It is also predicted that more than 50% of the research work related to fuzzy analysis will be revolving round fuzzy logic by 2025.

(iv) The most worked field related to fuzzy logic are concentrated to Computer Science and Engineering, which is evident enough by looking into the most prominent journals published in the same areas. However, it can be understood that as fuzzy logic has several interdisciplinary approach, thus it has tremendous applicability in vivid subject areas.

(v) China, Turkey, India, United States, Australia are the major producer of scientific records related to fuzzy logic and its applications in terms of documents as well as in terms of citations, co-authorship relationship, engaged Journals and institutions.

(vi) The renowned author working in fuzzy logic is Zadeh LA (father of fuzzy), whose involvement in this very area is commendable. Further, it also brings into notice that China is the most dynamic nation when it comes to fuzzy logic and its application as most of the researchers working for developing new ideas in fuzzy logic belong to China. United Kingdom is one of the top 10 highly productive nations representing European states standing out for publishing documents related to fuzzy analysis.

(vii) The three prominent journals that contributed toward eminent publications in the field of fuzzy logic are "Advances in Intelligent Systems and computing" with a count of 15 publications and 13 citations, followed by "Lecture Notes in Computer Science Including Subseries Lecture Notes in Artificial Intelligence and Lecture Notes in Bioinformatics", with a count of 14 publications and 19 citations and "Communications in Computer and Information Science" with a count of 12 publications but unfortunately no citations for it.

From the above analysis, it can be said that fuzzy logic and its application is the hotbed research area for the next generation, thereby creating new interrelated areas associated with it. However, due to the language barriers, keywords and database filters, a majority of research works were limited from above bibliometric analysis. Without these barriers, bibliometric analysis could have conducted over 2,35,014 records related to fuzzy analysis. However, for further development in the area of fuzzy analysis and applications other database, such as Elsevier, Scholar, Web of Science database can be taken into consideration for better interpretations and predictions.

References

1. Bellman, R.E. and Zadeh, L.A., Decision-Making in a Fuzzy Environment. *Manage. Sci.*, 17, 4, B–141, 1970, https://doi.org/10.1287/mnsc.17.4.B141.
2. Zadeh, L.A., Fuzzy sets. *Inf. Control*, 8, 3, 338–353, 1965, doi: https://doi.org/10.1016/S0019-9958(65)90241-X.
3. Dubois, D. and Prade, H., Fuzzy sets and systems: Theory and applications. *Math. Sci. Eng.*, 33, 2, p. 198, 1980, https://doi.org/10.1057/jors.1982.38.
4. Zadeh, L.A., Is there a need for fuzzy logic? *Inf. Sci.*, 178, 13, 2751–2779, 2008, https://doi.org/10.1016/j.ins.2008.02.012.
5. Blanco-Mesa, F., Merigó, J.M., Gil-Lafuente, A.M., Fuzzy decision making bibliometricbased review. *J. Intell. Fuzzy Syst.*, 32, 3, 2033–2050, 2017, https://doi.org/10.3233/JIFS-161640.
6. Merigó, J.M., Gil-Lafuente, A.M., Yager, R.R., An overview of fuzzy research with bibliometric indicators. *Appl. Soft Compu.*, 27, 420–433, 2015, https://doi.org/10.1016/j.asoc.2014.10.035.
7. Laengle, S., Lobos, V., Merigo, J.M., Herrera-Viedma, E., Cobo, M.J., De Baets, B., Forty years of fuzzy sets and systems: A bibliometric analysis. *Fuzzy Sets Syst.*, 402, 155–183, 2021, https://doi.org/10.1016/j.fss.2020.03.012.
8. Takagi, T. and Sugeno, M., Fuzzy identification of systems and its applications to modeling and control. *IEEE Trans. Syst. Man Cybern*, SMC-15, 1, 116–132, 1985, https://doi.org/10.1109/TSMC.1985.6313399.

9. Yu, D., Xu, Z., Wang, W., Bibliometric analysis of fuzzy theory research in China. *Knowl. Based Syst.*, 141, C, 188–199, 2018, https://doi.org/10.1016/j.knosys.2017.11.018.
10. López-Herrera, G. and M. J. C., *Visualization and evolution of the scientific structure of fuzzy sets research in Spain [Text]*, T.D. Wilson (Ed.), University of Boras, Sweden, 2009, Retrieved September 15, 2021, from http://informationr.net/ir/14-4/paper421.html.
11. Moher, D., Liberati, A., Tetzlaff, J., Altman, D.G., Group, T.P., Preferred Reporting Items for Systematic Reviews and Meta-Analyses: The PRISMA Statement. *PloS Med.*, 6, 7, e1000097, 2009, https://doi.org/10.1371/journal.pmed.1000097.
12. Cobo, M.J., López-Herrera, A.G., Herrera-Viedma, E., Herrera, F., Science mapping software tools: Review, analysis, and cooperative study among tools. *J. Am. Soc Inf. Sci. Technol.*, 62, 7, 1382–1402, 2011, https://doi.org/10.1002/asi.21525.
13. van Eck, N.J. and Waltman, L., Visualizing Bibliometric Networks, in: *Measuring scholarly impact: Methods and practice*, Y. Ding, R. Rousseau, D. Wolfram (Eds.), Cham, pp. 285–320, Springer International Publishing, 2014, https://doi.org/10.1007/978-3-319-10377-8_13.
14. Burnham, J.F., Scopus database review. *Biomed. Digit. Libr.*, 3, 1, 2006, https://doi.org/10.1186/1742-5581-3-1.
15. Van Eck, N.J. and Waltman, L., Text mining and visualization using VOSviewer, in: *ISSI Newsletter*, vol. 7, pp. 50–54.
16. Guo, Y.-M., Huang, Z.-L., Guo, J., Li, H., Guo, X., Nkeli, M., Bibliometric Analysis on Smart Cities Research. *Sustainability*, 11, 3606, 2019, https://doi.org/10.3390/su11133606.
17. Sevinc, A., Web of science: A unique method of cited reference searching. *J. Natl. Med. Assoc.*, 96, 7, 980–983, 2004.
18. Aytac, S., & Tran, C. Y. Authorship Trends of Research Articles Published in Seven Scientific, Technical, Engineering, and Medical (STEM) Library Journals: Analysis of STEM Library Research from 2016 to 2019 Update. Science & Technology Libraries, 40, 2, 117–132, 2021. https://doi.org/10.1080/0194262X.2021.1871702

5

Fuzzy Decision Making in Predictive Analytics and Resource Scheduling

Rekha A. Kulkarni[1*]**, Suhas H. Patil**[2] **and Bithika Bishesh**[3]

[1]*Dept. of Computer Engineering, SCTR's PICT, Pune, India*
[2]*Dept. of Computer Engineering, BVUCOE, Pune, India*
[3]*Department of Management, Sharda University, Greater Noida, India*

Abstract

Decision making can be defined as the process of selecting a particular solution from a set of alternatives by gathering the information and assessing the alternative solutions. It is a reasoning process based on assumptions of values, priorities, and objectives to be achieved. Effective decision making starts with identifying a query that needs a solution, getting insights into the probable solutions, and selecting the one that better fulfils objectives. While effective decision making heavily depends on the availability of relevant information, this information may have uncertainties in the real world scenario. Uncertainty may be in the form of vagueness in data because of the inherent nature of the real-world problems, which are based on degrees of truth rather than being completely true or false.

Approximate reasoning provides a way to work with imprecise or vague data while offering an acceptable solution to a real-world situation. Fuzzy logic plays a major role in this approximate reasoning as it resembles the human decision-making methodology. Fuzzy reasoning is an attractive approach to handling imprecise, unmodeled data in solving intelligent decision-making problems.

There are various application areas of fuzzy decision making like predictive analytics and scheduling of resources. This chapter gives an insight into fuzzy decision making and how it is different from traditional decision making. Handling of vagueness in data is discussed with the help of fuzzy logic, along with an overview of the application of fuzzy decision making in predictive analytics and resource scheduling.

***Keywords*:** Approximate reasoning, fuzzy decision tree, machine learning, resource scheduling, predictive analytics

**Corresponding author*: rakulkarni@pict.edu

Sachi Nandan Mohanty, Prasenjit Chatterjee and Bui Thanh Hung (eds.) Fuzzy Computing in Data Science: Applications and Challenges, (79–92) © 2023 Scrivener Publishing LLC

5.1 Introduction

Decision making is the process of making choices by identifying a problem, gathering information, and assessing alternative solutions. It is a reasoning process based on assumptions of values, priorities, and objectives to be achieved. Effective decision making starts with identifying a query that needs a solution, getting insights into the probable solutions and selecting the one that better fulfils the objectives. While effective decision making heavily depends on the availability of relevant information, this information may have uncertainties in the real-world scenario. Uncertainty may be in the form of vagueness in data because of the inherent nature of the real-world problems, which are based on degrees of truth rather than being completely true or false [3].

Approximate reasoning provides a way to work with imprecise or vague data while offering an acceptable solution to a real-world situation [2]. Fuzzy logic plays a major role in approximate reasoning as it resembles the human decision-making methodology. Fuzzy reasoning is an attractive approach to handling imprecise, unmodeled data in solving intelligent decision-making problems.

Fuzzy logic is a superset of conventional (Boolean) logic that has been extended to handle the concept of partial truth—truth values between "completely true" and "completely false." A type of logic that recognizes more than simple true and false values. With fuzzy logic, propositions can be represented with degrees of truthfulness and falsehood.

Fuzzy logic has proved to be particularly useful in expert systems and other artificial intelligence applications [5]. It is also used in some spell checkers to suggest a list of probable words to replace a misspelt one [10]. Fuzzy inference system (FIS) is the key unit of fuzzy logic. Fuzzy decision tree (FDT) combines fuzzy representation and approximate reasoning with symbolic decision trees. Decision trees are types of knowledge models driven by data representing a test on internal nodes and a decision on leaf nodes. Decision trees have become very popular tools for resource scheduling and predictive modeling used in various application areas of statistics, data mining and machine learning. Fuzzy decision trees can be employed in these fields while handling uncertainties and generating an acceptable solution.

Figure 5.1 shows an example of the difference between the conventional set theory and fuzzy logic.

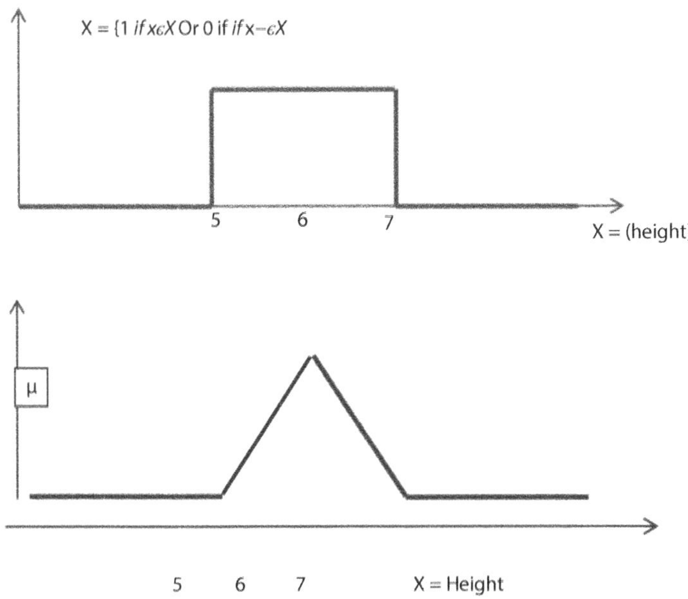

Figure 5.1 Difference between the conventional set theory and fuzzy logic.

5.2 History of Fuzzy Logic and Its Applications

Fuzzy logic is a new way of thinking about the world. First time the fuzzy logic was used was in 1965 when Lotfi Zadeh, a professor of electrical engineering and computer science at Berkeley, proposed it as an alternative to the binary system [1]. The concept was initially met with skepticism, but over time it has gained more acceptance as a way to deal with the uncertainties inherent in many real-world problems.

Fuzzy logic has been used in many different fields. It can be used for anything from controlling a thermostat to controlling traffic lights. Fuzzy logic is also being used in artificial intelligence, which will help computers better understand what people are saying and how they are feeling.

With the invention of machine learning and other supportive technologies, fuzzy logic has become one of the important aspects of building expert systems and bringing human-like decision-making power to computing systems.

Fuzzy logic can be applied to many real-life applications in various domain areas [4, 5, 10, 11].

- Aerospace: The altitude control and mixture regulation in deicing vehicles of spacecraft.
- Automotive: Trainable fuzzy systems for idle speed control, scheduling method for automatic transmission, intelligent highway systems, traffic control, improving the effectiveness of automatic transmissions.
- Business: Expert systems, decision-making support systems, personnel assessment systems.
- Chemical Industry: Control of pH, drying, chemical distillation processes, polymer extrusion production, oven gas cooling plant.
- Defence: Underwater object recognition, automatic target recognition of infrared images, naval decision support systems, control of hypervelocity interceptor, fuzzy set modeling of decision-making systems.
- Electronics: Control of automatic revelation in video cameras, humidity control, air conditioning systems, washing machine timers and home automation devices.
- Financial: Bank transaction control, ATM management, stock market predictions etc.
- Industrial: Heat exchanger control, wastewater treatment process control, water purification plant control, pattern analysis for industrial quality assurance, control of constraint satisfaction problems in structural design, control of water purification machines.
- Medical: Medical diagnostic support system, control of arterial pressure during anesthesia, multivariable control of anesthesia, modeling of neuropathological findings in Alzheimer's patients, radiology diagnoses, fuzzy inference diagnosis of diabetes and prostate cancer.
- Robotics: Fuzzy control for flexible-link manipulators, robot arm control.

5.3 Approximate Reasoning

Approximate reasoning is a method of information processing through Fuzzy rules [2]. These fuzzy rules are in the form of "if-then" constructs. For example the rule "if the size is big, then speed is slow," and we are given a (fuzzy) value for the size, then approximate reasoning gives us a method of computing a conclusion about the speed. The terms "big," "slow," and the data for "size" are all represented as fuzzy sets.

Approximate reasoning is concerned with evaluating a fuzzy rule like

if x is A, then y is B.

There are two universal sets X and Y from which x and y take their values, A and B are fuzzy subsets X (or Y). Fuzzy sets like A and B come from the values of linguistic variables. Consider two linguistic variables—size and speed. Size has members tiny, small, medium, large, very large, and huge. The members of size are defined by fuzzy membership functions. Fuzzy membership functions can be in the triangular or trapezoidal form [1].

Fuzzy logic works with four steps, fuzzification of the input variables, assigning membership function to individual input variables, and creating a rules database that goes as input to the inference engine. The inference engine acts as an expert decision-making system to arrive at an outcome based on several rules that have been fired for a given set of input variables. The fuzzy rule consists of many simple *if-then* rules of the form:

FR_i: if x is Ai and y is Bi and z is Ci, then O is S

where FR_i is the ith fuzzy rule and *x, y* and *z are* input variables, and O is the output (control) variable.

We are using the Mamdani-type rules because they provide a natural framework to include expert knowledge in linguistic rules.

5.4 Fuzzy Sets vs Classical Sets

Is a student regular in attending his classes? if the answer to such a question is in the form of "Yes" or "No," it is called a Set theory where the membership is either 1 or 0. But if the answer is in the form—a student is extremely regular, very regular, irregular, then we have the truth values associated with the membership, such as *partially true or partially false.*

In set theory which is also called crisp logic , an element will either completely belong to the set having common properties SA = {x:p(x)} or it will not belong to it:

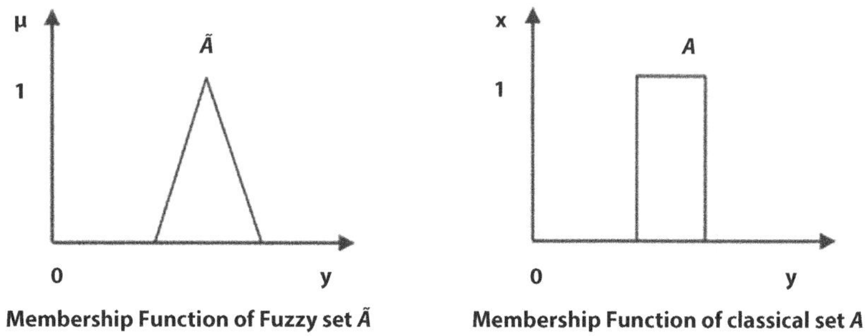

Figure 5.2 Comparison of the membership function of fuzzy sets with classical sets.

Example : The set SA= {1,3,5,7,9} is written as

$$SA = \{x: 1 \leq x < 10 \text{ and } (x\%2) \neq 0\}$$

Fuzzy sets can be considered as an extension and gross oversimplification of classical sets. Basically, it allows partial membership which means that it contains elements that have varying degrees of membership in the set [10].

Classical set contains elements that satisfy precise properties of membership while fuzzy set contains elements that satisfy imprecise properties of membership as given in equation 5.1.

$$\tilde{A} = \{(y, \mu_{\tilde{A}}(y)) \mid y = U\} \quad (5.1)$$

$(\mu_{\tilde{A}}(y))$ = Degree of membership of y which can have values from 0 to 1.

Figure 5.2 compares the membership function of a fuzzy set with the membership function of a classical set.

5.5 Fuzzy Inference System

Fuzzy Inference System (FIS) is the key unit of a fuzzy logic system having decision making as its primary work. It uses the "IF-THEN" rules along with connectors "OR" or "AND" for drawing essential decision rules [7].

5.5.1 Characteristics of FIS

Following are some characteristics of FIS:

- The output from FIS is always a fuzzy set irrespective of its input which can be fuzzy or crisp.
- It is necessary to have fuzzy output when it is used as a controller.
- A defuzzification unit would be there with FIS to convert fuzzy variables into crisp variables.
- Rule Base—It contains fuzzy IF-THEN rules.
- Database—It defines the membership functions of fuzzy sets used in fuzzy rules.
- Decision-making unit—It performs the operation on rules.
- Fuzzification Interface Unit—It converts the crisp quantities into fuzzy quantities.
- Defuzzification Interface Unit—It converts the fuzzy quantities into crisp quantities. Following is a block diagram of fuzzy interference system.

5.5.2 Working of FIS

The working of the FIS is shown in Figure 5.3. It consists of:

- Fuzzification unit—supports the application of numerous fuzzification methods, and converts the crisp input into fuzzy input.

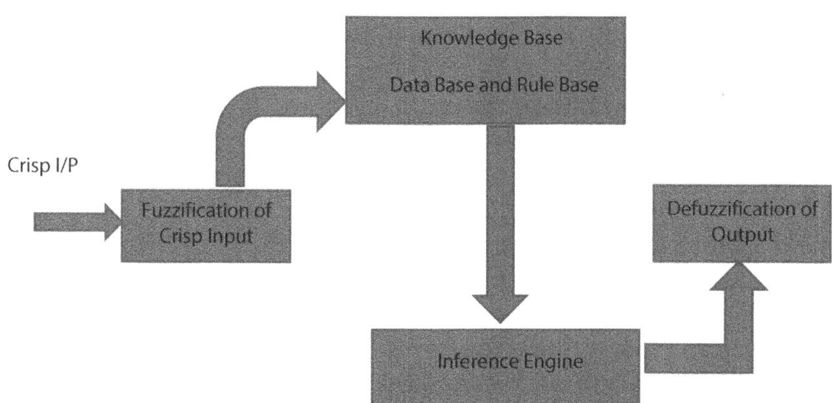

Figure 5.3 Fuzzy inference system.

- Knowledge base—forms collection of rule base and database upon the conversion of crisp input into fuzzy input.
- Defuzzification unit—fuzzy input is finally converted into crisp output.

5.5.3 Methods of FIS

The two important methods of FIS are [7]:

a. Mamdani Fuzzy Inference System- Mamdani-type inference provides the output in terms of membership function represented as fuzzy sets. The output variables so represented require defuzzification. By using the singleton output membership function, the efficiency of the defuzzification process can be increased as it simplifies the computations. The general Mamdani method uses the centroid method for the calculation of membership values. Instead of integrating across the two-dimensional function to find the centroid, the weighted average of a few data points is used.
b. Takagi-Sugeno Fuzzy Model (TS Method)—The Sugeno method of fuzzy inference system is quite similar to the Mamdani method. The first two steps of the inference process, fuzzifying the inputs and applying the fuzzy operator, are exactly the same. The main difference between Mamdani-type and Sugeno-type fuzzy inferences is that the output membership functions are only linear or constant for the Sugeno-type fuzzy inference.

5.6 Fuzzy Decision Trees

Fuzzy decision trees (FDTs) [10] are a generalization of decision trees to handle attributes with numeric-symbolic values, i.e., attributes associated either with a numerical value, for instance, the height is 172 cm or a symbolic value corresponding to a numerical measure, for instance, the height is tall. FDT is a very promising learning tool because it represents induced knowledge in a very expressive way [6]. The knowledge represented as an FDT is understandable and it differs from black-box systems as neural networks. Moreover, an FDT is equivalent to a set of fuzzy rules, and such kinds of induced rules can be introduced to optimize the query process of the database or to deduce decisions from data. FDTs enable us to obtain various kinds of such rules. An FDT, as a set of fuzzy rules, can be used as a knowledge base to apply in real life [9].

With the help of decision tree graphic models, text rules can be built and they are easily understood by the end user. Because of the inaccuracy of observations, and uncertainties, the data, collected in the environment, often take an unclear form [8]. Therefore, fuzzy decision trees are becoming popular in the field of machine learning. The knowledge acquired in the form of decision trees and inference procedures is highly valued for the clarity and visibility of the data. Initially, decision trees were popularized by Quinlan and his ID3 algorithm [7]. The fuzzy approach is becoming increasingly popular in solving problems of uncertainty, noise and inaccurate data. Nowadays, the fuzzy approach that integrates the concepts of fuzzy sets and entropy is becoming popular.

5.6.1 Characteristics of Decision Trees

Decision trees create effective and efficient models for machine learning. Following are the characteristics of decision trees:

- They are easily interpretable and visible;
- The model can be expressed both graphically and with text rules;
- Competitive in comparison with more expensive approaches;
- Decision trees are scalable;
- They can process discrete and continuous data;
- Decision trees can be applied to different sizes of data sets, including large sample sets.

5.6.2 Construction of Fuzzy Decision Trees

Define the fuzzy database, i.e., the fuzzy granulation for the domains of the continuous features.

1. Replace the continuous attributes of the training set using the linguistic.
2. labels of the fuzzy sets with the highest compatibility with the input values.
3. Calculate the entropy and information gain of each feature to split the training set and define the test nodes of the tree until all features are used or all training examples are classified.

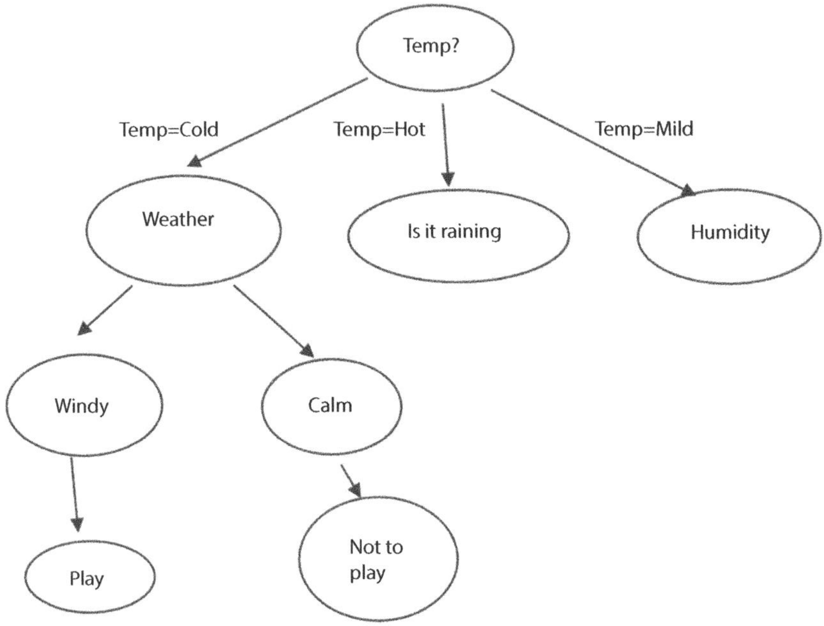

Figure 5.4 Showing partially constructed fuzzy decision tree.

Figure 5.4 shows an example of the fuzzy tree for deciding on whether to play or not to play based on weather conditions.

5.7 Fuzzy Logic as Applied to Resource Scheduling in a Cloud Environment

Virtual machine provisioning and migration services are an active part of the research which contribute to a self-adaptive and dynamic data centre, data scaling in private and public clouds, and high availability in clustered Virtual Machines (VMs) through live migration [13]. The performance improvement opportunities exist concerning the migration of VMs with V.M. Scheduling Algorithms and Accelerating VMs live migration time.

Real-time work-loads require the meeting of deadlines as a Quality of Service (QoS) parameter, and fuzzy logic can help us decide how to schedule the tasks on VMs and migrate the overloaded VM so that meeting the workflow deadlines can be improved. In our proposed system, we analyze the effect of using fuzzy logic on task mapping and migration service.

Table 5.1 shows an example of task sets and assigned fuzzy priorities to tasks. Based on the priorities, VMs will be allocated to tasks to ensure that the tasks' deadline is being met.

Table 5.2 shows the overall algorithm for our proposed system:

1. Input the workflow consisting of a set of tasks along with parameter values.
2. Fuzzify the input parameters execution time (E.T.), External priority (E.P.), and deadline (DL).
3. Based on the inference rules, calculate the overall priority.
4. Defuzzyficate the output, i.e., the overall priority.
5. Based on the priority, map the tasks to different types of VMs.
6. Check for a situation of an overloaded host.
7. Based on the information available at various hosts and with the help of fuzzy logic with minimum migration time to identify the host to Live, migrate the VMs.
8. Analyze if there are any deadline misses in the workflow.

Table 5.1 Tasks with parameters.

Task	Expected execution time EET	Assigned priority A.P.	Task deadline TDL	Fuzzy priority
T1	7	8	10	~10
T2	4	3	15	~12
T3	6	4	19	~14
T4	3	3	25	~17
T5	2	9	12	~9
T6	15	5	50	~20
T7	10	15	45	~18
T8	20	7	24	~25
T9	8	8	23	~24
T10	9	9	18	~23

Table 5.2 Assigning fuzzy membership values.

Fuzzy membership value	EET between 5 and 100	AP 1 to 10 (10 being highest)	Task deadline (10 to 50)
Low	$3 \leq$ EET< 20	$1\leq$ A.P. < 20	$1\leq$ T.D. < 20
Medium	$20\leq$EET<80	$20\leq$A.P. <80	$20\leq$T.D. <80
High	$80\leq$EET$\leq$100	$80\leq$AP$\leq$100	$80\leq$TD$\leq$100

Once the defuzzification of output parameters is done, mapping of tasks to a different type of VM will be performed. While performing the defuzzification, the method used is the center of gravity, as shown in equation 5.2.

$$COG = \frac{\sum_{i=1}^{m} \mu A(x)x}{\sum_{i=1}^{m} \mu A(x)} \tag{5.2}$$

In our algorithm, we consider the minimum migration time as the policy for migrating the V.M. from an overloaded Host. In identifying the overload, we use the threshold such that CPU utilization of VM_a is less than the CPU threshold as given in equation 5.3.

$$CPU_{utilization} \leq CPU_{threshold} \tag{5.3}$$

So, this method compares the migration time and selects the VM_a with a minimum migration time among all VMs in host has per equation 5.4.

$$a \in V_h \mid \forall\, b \in V_h, \frac{RAM(a)}{NETh} \leq \frac{RAM(b)}{NETh} \tag{5.4}$$

5.8 Conclusion

Fuzzy decision making helps in handling the uncertainties in the input values [12]. Scheduling problems with the fuzzy mapping of resources help give better results for real-time scheduling of workflows. Fuzzy logic gives better results when used for predictive analysis with smaller datasets. The performance of Fuzzy decision making can be improvised with some optimization techniques to map the membership functions.

References

1. Zadeh, L.A., Fuzzy sets. *Inf. Control*, 8, 394–432, 1965.
2. Buckley, J.J. and Eslami, E., Approximate Reasoning, in: *An Introduction to Fuzzy Logic and Fuzzy Sets*, pp. 231–252, Physica, Heidelberg, 2002.
3. Tchernykh, A., Schwiegelsohn, U., Alexandrov, V., Talbi, E.G., Towards understanding uncertainty in cloud computing resource provisioning. *Proc. Comput. Sci.*, *51*, 1772–1781, 2015.
4. Costea, A., Applying fuzzy logic and machine learning techniques in financial performance predictions. *Proc. Econom. Financ.*, *10*, 4–9, 2014.
5. Chopra, P. and Bedi, R., Applications of fuzzy logic in cloud computing: A review. *Int. J. Sci. Res. Eng. Technol.*, 6, 1083–1086, 2017.
6. Begenova, S.B. and Avdeenko, T.V., Building of fuzzy decision trees using ID3 algorithm, in: *Journal of Physics: Conference Series*, vol. 1015, IOP Publishing, p. 022002, 2018, May.
7. LaCasse, P.M., Otieno, W., Maturana, F.P., A hierarchical, fuzzy inference approach to data filtration and feature prioritization in the connected manufacturing enterprise. *J. Big Data*, 5, 1, 1–31, 2018.
8. Janikow, C.Z., Fuzzy decision trees: issues and methods. *IEEE Trans. Syst. Man Cybern. Part B (Cybern.)*, *28*, 1, 1–14, 1998.
9. Chen, Y.L., Wang, T., Wang, B.S., Li, Z.J., A survey of fuzzy decision tree classifier. *Fuzzy Inf. Eng.*, *1*, 2, 149–159, 2009.
10. Ross, T.J., *Fuzzy logic with engineering applications*, John Wiley & Sons, West Sussex, PO19 8SQ, United Kingdom, 2005.
11. Rajasekaran, S. and Pai, G.V., *Neural networks, fuzzy logic and genetic algorithm: Synthesis and applications (with cd)*, PHI Learning Pvt. Ltd, Delhi, India, 2003.
12. Shah, C., *A hands-on introduction to data science*, Cambridge University Press, Cambridge CB2 8BS, United Kingdom, 2020.
13. Monil, M.A.H. and Rahman, R.M., VM consolidation approach is based on heuristics, fuzzy logic, and migration control. *J. Cloud Comput.*, *5*, 1, 1–18, 2016.

6

Application of Fuzzy Logic and Machine Learning Concept in Sales Data Forecasting Decision Analytics Using ARIMA Model

S. Mala* and V. Umadevi

Department of Computer Science, Nehru Memorial College (Affiliated to Bharathidasan University), Tiruchirapalli, India

Abstract

This chapter focuses on prediction-based sales data analysis with four types of seeds sample, such as spinach, black gram, gingelly, and fodder sorghum. The proposed work has applied a fuzzy logic technique, namely clustering and artificial intelligence algorithms for evaluating sales variety of seeds and period, and the performance of sales in a particular place has also been analyzed. The sample data collected and forecasting timeseries employed Auto-Regressive Integrated Moving Average (ARIMA model). Both the comparisons of Sales Place and Seeds Variety and Sales Period (Seasons) have been carried out for the prediction analysis work. The chapter concluded that forecasting sales values in terms of time helps to identify the required quantity of purchase and leads to avoid the unwanted loss and wastage of seeds. The aim of the present work assisted is to investigate sales trends for a given time period and to build optimal sales model based on season and location. The prediction analysis finds the sales behavior across time period estimated by R-Tool. In this chapter, the work in the quantity of sales prediction for different types of seeds is done based on season and location. The sale on time can prevent wastage of seeds and increase profits, based not only on time but also on location.

***Keywords*:** Fuzzy logic, machine learning, ARIMA model, optimal model, forecasting analysis, time series analysis

**Corresponding author*: ursmala@gmail.com

Sachi Nandan Mohanty, Prasenjit Chatterjee and Bui Thanh Hung (eds.) Fuzzy Computing in Data Science: Applications and Challenges, (93–112) © 2023 Scrivener Publishing LLC

6.1 Introduction

This chapter focuses on prediction-based sales data analysis with four types of seeds sample, such as spinach, black gram, gingelly, and fodder sorghum. The sample data that have been collected and forecasting time series has employed Auto-Regressive Integrated Moving Average (ARIMA MODEL). Both the comparisons of Sales Place and Seeds Variety and also Sales Period (Seasons) have taken for the prediction analysis work. This chapter concluded with forecasting sales values in terms of time that helps to identify the required quantity of purchase and leads to avoid the unwanted loss and wastage of seeds. The aim of the present work assisted to investigate sales trends for given time period and to build optimal sales model. The prediction analysis finds the sales behavior across time period estimated by R-Tool.

6.1.1 Aim and Scope

- To compare the sales data of various seed using optimization,
- To predict the suitable seed sale in particular season,
- To categorize the seeds, seasons and sales of analysis for forecasting,
- To build the ARIMA model for forecasting analysis.

6.1.2 R-Tool

R-It is a software programming language and software environment for statistical computing and graphics. It is an integrated suite of software facilities for data manipulation, calculation, and graphical display. It is an open source, which includes virtually every data manipulation techniques, statistical model, and chart. Conventional spreadsheet applications, like Excel, are not able to handle large-scale datasets. We plan to use R primarily for preprocessing the data, manipulating it into structured form for the analysis and present the end results.

6.1.3 Application of Fuzzy Logic

The review analyzed typical applications and highlights potential contributions that fuzzy set theory can make to machine learning, data mining, and related fields. The chapter concludes with a critical consideration of recent developments and some suggestions for future research directions.

The functional relation between the data points and the coefficients of a linear regression function can be extended to the case of fuzzy data, where the observations are described in terms of fuzzy sets.

The class performance variable is obtained by applying on the performance dataset the Fuzzy data set and obtaining clusters with similar performance. We attach to each input dataset observation a performance class, depending on which cluster contains the observation given the characterization and hierarchy of the sales of place, period and variety of seed, in "good," "medium," and "poor" performance.

6.1.4 Dataset

AVIS seeds company sales data are taken (Figure 6.1). The company was started in 2018. The company seeds provided to 32 shops in 25 places. The company is having four type of seeds namely Spinach, black gram, gingelly, fodder sorghum, and six types of Spinach, namely mulaikeerai, thandukeerai, arakeerai, pulichakeerai, sirukeerai, palakeerai, and three types of black gram as follows black gram VBN4, black gram VBN5, black gram VBN6, and two types of gingelly, gingelly TMV4 and gingelly TMV7. The data entry is done by month wise, and it is converted every 15 days. The Seeds Sales Dataset is in structured format. The dataset has 7 attributes and 192 rows.

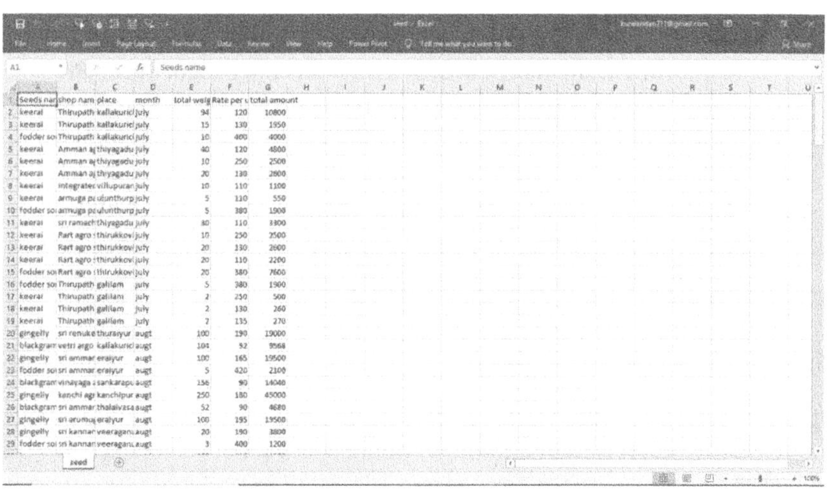

	Seeds nar	shop nam	place	month	total weig	Rate per u	total amount
2	keerai	Thirupath	kallakurici	july	94	120	10800
3	keerai	Thirupath	kallakurici	july	15	130	1950
4	fodder so	Thirupath	kallakurici	july	10	400	4000
5	keerai	Amman a	thiyagadu	july	40	120	4800
6	keerai	Amman a	thiyagadu	july	10	250	2500
7	keerai	Amman a	thiyagadu	july	20	130	2800
8	keerai	integratec	villupuran	july	10	110	1100
9	keerai	armuga p	ulunthurp	july	5	110	550
10	fodder so	armuga p	ulunthurp	july	5	380	1900
11	keerai	sri ramach	thiyagadu	july	30	110	3300
12	keerai	Rart agro	thirukkovi	july	10	250	2500
13	keerai	Rart agro	thirukkovi	july	20	130	2600
14	keerai	Rart agro	thirukkovi	july	20	110	2200
15	fodder so	Rart agro	thirukkovi	july	20	380	7600
16	fodder so	Thirupath	galilam	july	5	380	1900
17	keerai	Thirupath	galilam	july	2	250	500
18	keerai	Thirupath	galilam	july	2	130	260
19	keerai	Thirupath	galilam	july	2	135	270
20	gingelly	sri renuke	thuraiyur	augt	100	190	19000
21	blackgram	vetri argo	kallakurici	augt	104	92	9568
22	gingelly	sri ammar	eraiyur	augt	100	165	19500
23	fodder so	sri ammar	eraiyur	augt	5	420	2100
24	blackgram	vinayaga	sankarapu	augt	156	90	14040
25	gingelly	kanchi ag	kanchipur	augt	250	180	45000
26	blackgram	sri ammar	thalaivasa	augt	52	90	4680
27	gingelly	sri arumu	eraiyur	augt	100	195	19500
28	gingelly	sri kannar	veeragan	augt	20	190	3800
29	fodder so	sri kannar	veeragan	augt	3	400	1200

Figure 6.1 Sample seed dataset.

6.2 Model Study

6.2.1 Introduction to Machine Learning Method

Machine learning is an application of artificial intelligence (AI) that provides systems the ability to automatically learn and improve from experience without being explicitly programmed. Machine learning focuses on the development of computer programs that can access data and use it to learn for themselves. The process of learning begins with observations or data, such as examples, direct experience, or instruction, in order to look for patterns in data and make better decisions in the future based on the examples that we provide. The primary aim is to allow the computers to learn automatically without human intervention or assistance and adjust actions accordingly.

Machine learning algorithms are often categorized as supervised or unsupervised.

- Supervised machine learning algorithms can apply what has been learned in the past to new data using labeled examples to predict future events. Starting from the analysis of a known training dataset, the learning algorithm produces an inferred function to make predictions about the output values. The system is able to provide targets for any new input after sufficient training. The learning algorithm can also compare its output with the correct, intended output and find errors in order to modify the model accordingly.
- In contrast, unsupervised machine learning algorithms are used when the information used to train is neither classified nor labeled. Unsupervised learning studies how systems can infer a function to describe a hidden structure from unlabeled data. The system does not figure out the right output, but it explores the data and can draw inferences from datasets to describe hidden structures from unlabeled data.

6.2.2 Time Series Analysis

A time series is a sequential set of data points, measured typically over successive times. It is mathematically defined as a set of vectors x(t), t = 0,1, 2... where t represents the time elapsed [1–3]. The variable x (t is treated as a random variable). The measurements taken during an event in a time series are arranged in a proper chronological order. A time series containing records of a single variable is termed as univariate. But if records of more than one variable are considered, it is termed as multivariate. A time series can be continuous or discrete. In a continuous time, series observations are measured at every instance of time, whereas a discrete time series contains observations measured at discrete points of time. For example, temperature readings, flow of a river, concentration of a chemical process, etc., can be recorded as a continuous time series. On the other hand, the population of a particular city, production of a company, exchange rates between two different currencies may represent discrete time series. Usually, in a discrete time series, the consecutive observations are recorded at equally spaced time intervals, such as hourly, daily, weekly, monthly, or yearly time separations. As mentioned in [5], the variable being observed in a discrete time series is assumed to be measured as a continuous variable using the real number scale.

6.2.3 Components of a Time Series

A time series in general is supposed to be affected by four main components, which can be separated from the observed data. These components are as follows: trend, cyclical, seasonal, and irregular components. A brief description of these four components is given here. The general tendency of a time series to increase, decrease or stagnate over a long period of time is termed as secular trend or simply trend. Thus, it can be said that a trend is a long-term movement in a time series. For example, series relating to population growth, number of houses in a city etc. show upward trend, whereas downward trend can be observed in series relating to mortality rates, epidemics, etc. Seasonal variations in a time series are fluctuations within a year during the season. The important factors causing seasonal variations are: climate and weather conditions, customs, traditional habits, etc. For example, sales of ice-cream increase in summer, sales of woollen cloths increase in winter. Seasonal variation is an important factor for businessmen, shopkeeper, and producers for making proper future plans [6]. The cyclical variation in a time series describes the medium-term changes in the series, caused by circumstances, which repeat in cycles. The duration of a cycle extends over longer period of time, usually 2 or more years. Most of the economic and financial time series show some kind of cyclical variation. For example, a business cycle consists of four phases, *viz.*

i) Prosperity, ii) Decline, iii) Depression and iv) Recovery.

Schematically a typical business cycle can be shown as below:

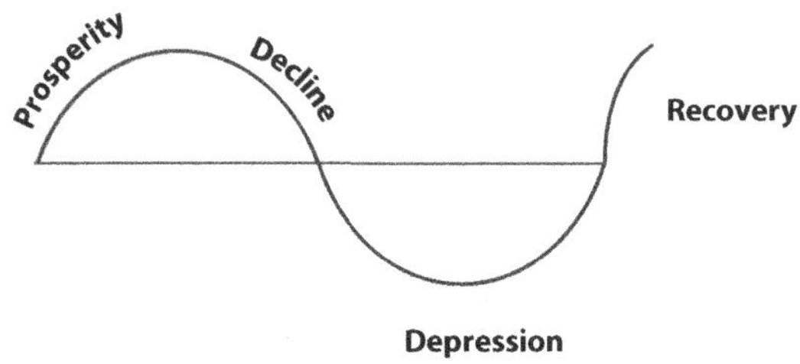

Figure 6.2 A four-phase business cycle.

In four-phase business cycle, 6 irregular or random variations in a time series are caused by unpredictable influences, which are not regular and also do not repeat in a particular pattern (Figure 6.2). These variations are caused by incidences, such as war, strike, earthquake, flood, revolution, etc. There is no defined statistical technique for measuring random fluctuations in a time series. Considering the effects of these four components, two different types of models are generally used for a time series *viz.* multiplicative and additive models. Multiplicative model: Y(t) = T(t)× S(t)×C(t)× I additive model: Y(t) = T(t) + S(t) + C(t) + I. Here, Y(t) is the observation and T(t), S(t) C(t), and I(t) are respectively the trend, seasonal, cyclical, and irregular variation at time, t. Multiplicative model is based on the assumption that the four components of a time series are not necessarily independent, and they can affect one another, whereas in the additive model, it is assumed that the four components are independent of each other. The concept of stationary of a stochastic process can be visualized as a form of statist Furthermore, a continuous time series can be easily transformed to a discrete one by merging data together over a specified time interval.

$$\text{Multiplicative Model: } Y(t) = T(t)\times S(t)\times C(t)\times I$$

$$\text{Additive Model: } Y(t)=T(t)+S(t)+C(t)+ I$$

Here, Y(t) is the observation and T(t), S(t) C(t) and I(t) are respectively the trend, seasonal, cyclical and irregular variation at time, t. Multiplicative model is based on the assumption that the four components of a time series are not necessarily independent and they can affect one another, whereas in the additive model, it is assumed that the four components are independent of each other.

6.2.4 Concepts of Stationary

The concept of stationary of a stochastic process can be visualized as a form of statistical equilibrium [4]. The statistical properties, such as mean and variance of a stationary process, do not depend upon time. It is a necessary condition for building a time series model that useful for future forecasting. Furthermore, the mathematical complexity of the fitted model reduces with this assumption. There are two types of stationary processes which are defined below process {x(t),t=0,1,2...} is strongly stationary or strictly stationary if the joint probability distribution function of {xt−s,xt−s+1...,xt... xt+s−1, xt+s} is independent of for alls. Thus, for a strong stationary process, the joint distribution of any possible set of random variables from the processes independent of time [1, 3].

However, for practical applications, the assumption of strong stationarity is not always needed and so a somewhat weaker form is considered. A stochastic process is said to be Weakly Stationary of order k if the statistical moments of the process up to that order depend only on time differences and not upon the time of occurrences of the data being used to estimate the moments [1–3]. For example, a stochastic process $\{x(t), t = 0,1, 2...\}$ is second-order stationary [1, 3] if it has time-independent mean and variance and the covariance values *Cov*(xt, xt−s) depend only ones.

It is important to note that neither strong nor weak stationarity implies the other. However, a weakly stationary process following normal distribution is also strongly stationary [1]. Some mathematical tests like the one given by Dickey and Fuller [5] are generally used to detect stationarity in a time series data.

As mentioned in [5, 6], the concept of stationarity is a mathematical idea constructed to simplify the theoretical and practical development of stochastic processes. To design a proper model, adequate for future forecasting, the underlying timeseries is expected to be stationary. Unfortunately, it is not always the case. As stated by Hipel and McLeod [6], the greater the timespan of historical observations, the greater is the chance that the time series will exhibit nonstationary characteristics. However, for a relatively short time span, one can reasonably model the series using a stationary stochastic process. Usually, a time series showing trend or seasonal patterns is nonstationary in nature. In such cases, differencing and power transformations are often used to remove the trend and to make the series stationary.

6.2.5 Model Parsimony

While building a proper time series model, we have to consider the principle of parsimony [2–4]. According to this principle, always the model with smallest possible number while building a proper time series model, we have to consider the principle of parsimony of parameters is to be selected so as to provide an adequate representation of the underlying time series data [2]. Out of a number of suitable models, one should consider the simplest one, still maintaining an accurate description of inherent properties of the time series. The idea of model parsimony is similar to the famous Occam's razor principle [4]. As discussed by Hipel and McLeod [6], one aspect of this principle is that when faced with a number of competing and adequate explanations, pick the simple stone. The Occam's razor provides considerable inherent information's, when applied to logical analysis.

Moreover, the more complicated the model, the more possibilities will arise for departure from the actual model assumptions. With the increase of model parameters, the risk of overfitting also subsequently increases. An overfitted timeseries model may describe the training data very well, but it may not be suitable for future forecasting. As potential overfitting affects the ability of a model to forecast well, parsimony is often used as guiding principle to overcome this issue. Thus, in summary, it can be said that, while making timeseries forecasts, genuine attention should be given to select the most parsimonious model among all other possibilities.

6.3 Methodology

6.3.1 Exploratory Data Analysis

EDA is an approach to analysing data sets to summarize their main characteristics, often with visual methods. A statistical model can be used or not, but primarily, EDA is for seeing what the data can tell us beyond the formal modelling or hypothesis testing task. Exploratory data analysis was promoted by John Tukey to encourage statisticians to explore the data, and possibly formulate hypotheses that could lead to new data collection and experiments. EDA is different from initial data analysis (IDA), which focuses more narrowly on checking assumptions required for model fitting and hypothesis testing, and handling missing values and making transformations of variables as needed.

6.3.1.1 *Seed Types—Analysis*

With the above comparison of seeds (Table 6.1), it can be concluded that the maximum sales happen in black gram, and the minimum seeds sale is in fodder sorghum. Black gram 50% sales. Fodder Sorghum, 2% sales; gingelly, 39% sales; spinach, 9% sales (Figure 6.3).

6.3.1.2 *Comparison of Location and Seeds*

With the abovementioned comparison of the place and seeds (Table 6.2), it can be concluded that the maximum sales happen in Ariyamangalam and the minimum seeds sale is in Omalur. Black gram seeds recorded 1324 kg

Table 6.1 Seed types.

Seed type	Sum of total weight (in gms)
Blackgram	7692
Fodder sorghum	280
Gingelly	5906
Spinach	1423
Grand Total	**15301**

Figure 6.3 Seed sales data.

Table 6.2 Location-based seed sales.

Location/seeds	**Black gram**	**Fodder sorghum**	**Gingelly**	**Spinach**	**Grand total**
Ariyamangalam	**1324**	**75**	**650**	--	**2049**
Atthur	**40**	--	--	--	**40**
Dharapuram		20		**20**	**40**
Eraiyur	**240**	**5**	**450**	**60**	**755**
Erode	**20**	450	33		503
Galilam		5		**6**	**11**
Kallakurichi	768	30	540	557	1895
Kanchipuram	920	25	700	69	1714
Kirshnagiri	**500**	**10**			510
Madurai	312		50	45	407
Omalur				**10**	**10**
Panruti		20	700	35	755
Rasipuram				250	250

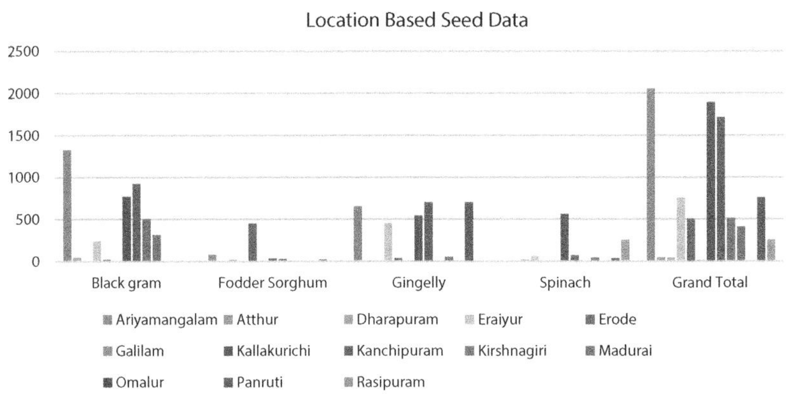

Figure 6.4 Location-based seed data.

Table 6.3 Season-based seed data.

Month/ seeds	Black gram	Fodder sorghum	Gingelly	Spinach	Grand total
Jan	1300	20	950	145	2415
Feb	208	125	1230	306	1869
July		40		280	320
Oct	4244	22		86	4352
Nov	572		1126	105	1803
Dec	504	30	1480	266	2280
Augt	864	8	1120	150	2142
Sept		35		85	120

sales in ariyamangalam and fodder sorghum seeds about 75 kg sale in ariyamangalam. 800 kg of gingelly seeds has been sold in thiyagadhurgam. 557 kg of spinach seeds has been sold in Kallackurichi (Figure 6.4).

6.3.1.3 Comparison of Season (Month) and Seeds

With the abovementioned comparison of month and seeds (Table 6.3), it can be concluded that seeds that the maximum sale happens in October month and the minimum seeds sale is in September. Black gram seeds recorded 4244 kg sale in October month and Fodder Sorghum Seeds about 125 kg Sale in February month. 1480 kg of Gingelly Seeds has been sold in December month. 306 kg of spinach seeds has been sold in February month.

6.3.2 Forecasting

6.3.2.1 Auto Regressive Integrated Moving Average (ARIMA)

Auto Regressive Integrated Moving Average (ARIMA) models are, in theory, the most general class of models for forecasting a time series which can be made to be "stationary" by differencing (if necessary), perhaps in conjunction with nonlinear transformations, such as logging or deflating (if necessary). A random variable that is a time series is stationary if its statistical properties are all constant over time. A stationary series has no trend, its variations around its mean have a constant amplitude, and it wiggles in a consistent fashion, i.e., its short-term random time patterns always look the same in a statistical sense. The latter condition means that its *autocorrelations* (correlations with its own prior deviations from the mean) remain constant over time, or equivalently, that its power spectrum remains constant over time. A random variable of this form can be viewed (as usual) as a combination of signal and noise, and the signal (if one is apparent) could be a pattern of fast or slow mean reversion, or sinusoidal oscillation, or rapid alternation in sign, and it could also have a seasonal component. An ARIMA model can be viewed as a "filter" that tries to separate the signal from the noise, and the signal is then extrapolated into the future to obtain forecasts.

The ARIMA forecasting equation for a stationary time series is a linear (i.e., regression-type) equation in which the predictors consist of lags of the dependent variable and/or lags of the forecast errors. That is:

Predicted value of Y = a constant and/or a weighted sum of one or more recent values of Y and/or a weighted sum of one or more recent values of the errors.

If the predictors consist only of lagged values of Y, it is a pure autoregressive ("self-regressed") model, which is just a special case of a regression model and which could be fitted with standard regression software. For example, a first-order autoregressive ("AR (1)") model for Y is a simple regression model in which the independent variable is just Y lagged by one period (LAG(Y,1) in Stat graphics or Y_LAG1 in Regress It). If some of the predictors are lags of the errors, an ARIMA model it is NOT a linear regression model, because there is no way to specify "last period's error" as an independent variable: the errors must be computed on a period-to-period basis when the model is fitted to the data. From a technical standpoint, the problem with using lagged errors as predictors is that the model's predictions are not linear functions of the coefficients, even though they are linear functions of the past data. So, coefficients in ARIMA models that include lagged errors must be estimated by nonlinear optimization methods ("hill-climbing") rather than by just solving a system of equations.

The acronym ARIMA stands for Auto-Regressive Integrated Moving Average. Lags of the stationaries series in the forecasting equation are called "autoregressive" terms, lags of the forecast errors are called "moving average" terms, and a time series which needs to be differenced to be made stationary is said to be an "integrated" version of a stationary series. Random-walk and random-trend models, autoregressive models, and exponential smoothing models are all special cases of ARIMA models.

A nonseasonal ARIMA model is classified as an "ARIMA (p, d, q)" model, where:

- p is the number of autoregressive terms,
- d is the number of nonseasonal differences needed for stationarity, and
- q is the number of lagged forecast errors in the prediction equation.

The forecasting equation is constructed as follows. First, let y denote the dth difference of Y, which means:

$$\text{If } d=0: yt = Yt$$

$$\text{If } d=1: yt = Yt - Yt\text{-}1$$

$$\text{If } d=2: yt = (Yt - Yt\text{-}1) - (Yt\text{-}1 - Yt\text{-}2) = Yt - 2Yt\text{-}1 + Yt\text{-}2$$

Note that the second difference of Y (the d=2 case) is not the difference from two periods ago. Rather, it is the first-difference-of-the-first difference, which is the discrete analog of a second derivative, i.e., the local acceleration of the series rather than its local trend.

In terms of y, the general forecasting equation is:

$$\hat{y}t = \mu + \phi 1\ yt\text{-}1 + \ldots + \phi pyt\text{-}p - \theta 1et\text{-}1 - \ldots - \theta qet\text{-}q$$

Here, the moving average parameters (θ's) are defined so that their signs are negative in the equation, following the convention introduced by Box and Jenkins. Some authors and software (including the R programming language) define them so that they have plus signs instead. When actual numbers are plugged into the equation, there is no ambiguity, but it is important to know which convention your software uses when you are reading the output. Often the parameters are denoted there by AR (1), AR (2), ..., and MA (1), MA (2), etc, to identify the appropriate ARIMA model for Y, you begin by determining the order of differencing (d), needing to stationary the series and remove the gross features of seasonality, perhaps in conjunction with a variance-stabilizing transformation, such as logging or deflating. If you stop at this point and predict that the differenced series is constant, you have merely fitted a random walk or random trend model. However, the stationaries series may still have autocorrelated errors, suggesting that some number of AR terms ($p \geq 1$) and/or some number MA terms ($q \geq 1$) are also needed in the forecasting equation.

6.3.2.2 Data Visualization

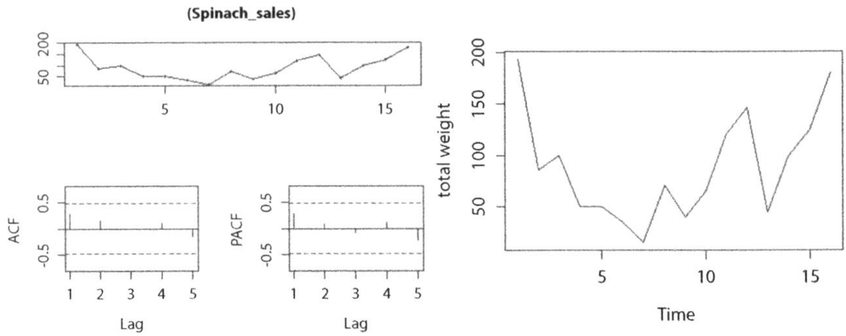

Figure 6.5 Spinach sales.

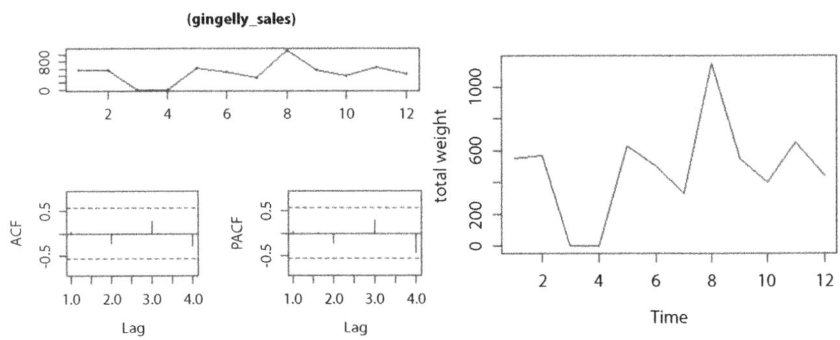

Figure 6.6 Gingelly sales.

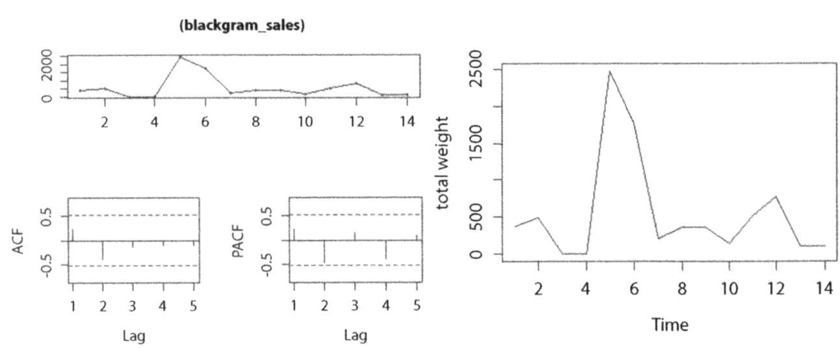

Figure 7 Black gram sales.

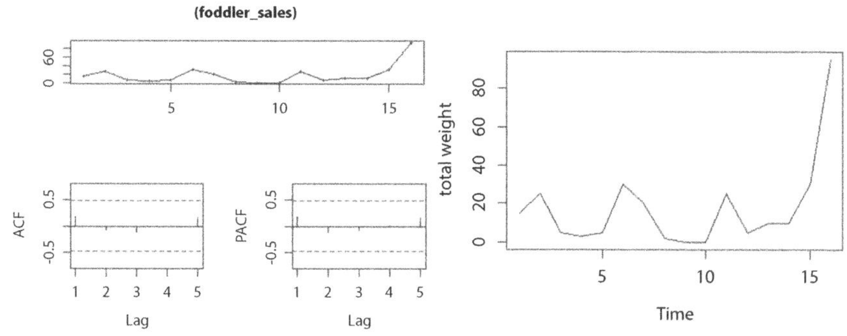

Figure 6.8 Fodder sorghum sales.

Data: Spinach sales Dickey-Fuller = -2.2203, Lag order = 2, p-value = 0.4884
alternative hypothesis: stationary (Figure 6.5)
Data: Gingelly sales Dickey-Fuller = -0.98536, Lag order = 2, p-value = 0.9227
alternative hypothesis: stationary (Figure 6.6)
Data: Blackgram Dickey-Fuller = -2.1255, Lag order = 2, p-value = 0.5246
alternative hypothesis: stationary (Figure 6.7)
Data: Fodder sorghum Dickey-Fuller = -0.39983, Lag order = 2, p-value = 0.9793
alternative hypothesis: stationary (Figure 6.8)

All the seeds data are stationary. Reason is: H0 Then unit root for the series. H1 Then no unit root for the series, (the series stationary) α= values are 0.05%

> $\alpha \geq p$ Accept null hypothesis and reject alternative hypothesis
> $\alpha \leq p$ rejects null hypothesis and accept alternative hypothesis (this is stationary).

All the autocorrelation function (ACF) and partial autocorrelation (PACF) value is between −0.5 and 0.5 so the data are white noise data. There is no correlation between the random variables, and the mean value is zero. There is no multicollinearity between the data. So, ARIMA function, but alternative models are there. Any time series models can be predicted.

(1,0,0) First-order autoregressive model
(0,1,0) Random walk
(1,1,0) Differenced first-order autoregressive model
(0,1,1) Without constant = simple exponential smoothing
(0,1,1) with constant = simple exponential smoothing with growth
(0,2,1) or (0,2,2) without constant = linear exponential smoothing
(1,1,2) without constant = damped-trend linear exponential smoothing

6.3.2.3 *Implementation Model*

The data are white noise data. There is no correlation between the random variables, and mean value is zero. There is no multicollinearity between the data, so, ARIMA function, but alternative models are there. Any time series models can be predicted.

The seeds data set fits only for two models.

- without constant = linear exponential smoothing (0,2,2)
- without constant = damped-trend linear exponential smoothing (1,1,2)

6.4 Result Analysis

Black gram dataset prediction for every next 15 days. The data are predicted six times. So, the sales weight predicting is done for three months (Figure 6.9).

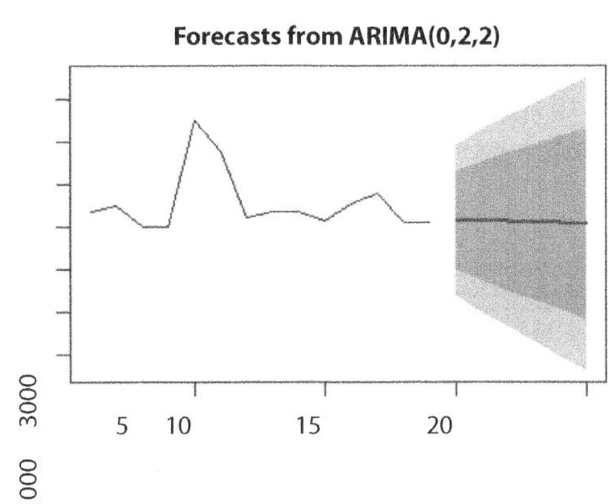

Figure 6.9 Prediction for black gram.

Spinach dataset prediction for every 15 days. The data are predicted six times. So, the sales weight predicting is done for three months (Figure 6.10).

Gingelly dataset prediction for every next 15 days. The data are predicted six times. So, the sales weight predicting is done for three months (Figure 6.11).

Fodder sorghum dataset prediction for every next 15 days. The data are predicted six times. So, the sales weight predicting is done for 3 months (Figure 6.12).

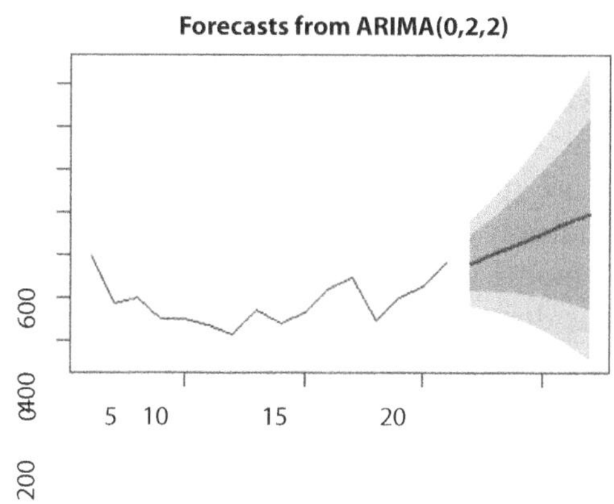

Figure 6.10 Prediction for spinach dataset.

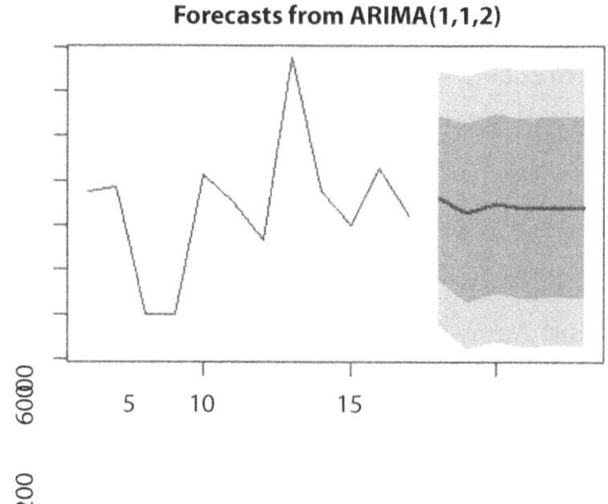

Figure 6.11 Prediction for gingelly dataset.

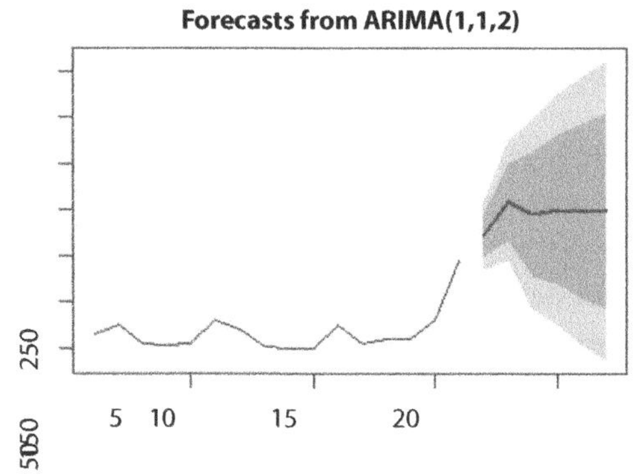

Figure 6.12 Prediction for fodder dataset.

6.5 Conclusion

The data divided are spinach, black gram, gingelly, fodder sorghum, and the duration of monthly data is converted into every 15 days. The reason that the ARIMA model works depending upon the time series. The prediction of data has done six times hence the data prediction is for next 3 months. Gingelly seeds mostly sell on the first 15 days of the first month. Fodder sorghum seeds mostly sell on last 15 days of first month. Spinach seeds mostly sell on last 15 days of third month. Black gram seeds mostly sell on first 15 days of first month. The next 3 months gingelly seeds will be the most sold ones. The total weight of gingelly seeds is 2894 kg. The total weight of seeds fodder sorghum 870 kg. The total weight of seeds spinach is 1419 kg. The total weight of seeds black gram is 723 kg. The lowest sale made is black gram seeds. Thus, in the future, this project can be used in various other domains to predict the sales and other criteria. The applicable way from the forecasting analysis with R-Tool the seller can predict the time period of sales of seeds and capacity.

References

1. Adhikari, R. and Agrawal, R.K., *An introductory study on time series modeling and forecasting*, pp. 18–23, LAP Lambert Academic Publishing, Germany, 2013, http://arxiv.org/ftp/arxiv/papers/1302/1302.6613.pdf.
2. Box, G.E.P. and Jenkins, G.M., *Time series analysis: Forecasting and control, revised edition, holden-day*, John Wiley & Sons Inc., Hoboken, New Jersey, 1976, https://www.math.kth.se/matstat/gru/sf2943/ts.pdf.

3. Brockwell, P.J. and Davis, R.A., *Time series: Theory and methods*, 2nd Edition, Springer-Verlag, New York, 1991, https://www.statisticssolutions.com/time-series-analysis/.
4. Jenkins Methodology, in: *Applied econometrics*, 2nd, pp. 265–286, Palgrave MacMillan, Palgrave, UK, 2016, ISBN: 978-0-230-27182-1, https://people.duke.edu/~rnau/411arim.htm.
5. Sanz-Fernandez, V., Cabrera, R., Munoz-Lechunga, R., Sanchez-Navas, A., Czerwinski, I., Development of medelling script of time series suitable for data mining. *Open J. Stat.*, 6, 555–564, 2016.
6. Hipel, K.W. and MCLeod, A.I., *Time series modelling of water resources and environmental systems*, 1st Edition, Elsevier, Amsterdam, April 7, 1994.

7

Modified m-Polar Fuzzy Set ELECTRE-I Approach

Madan Jagtap[1*], Prasad Karande[2] and Pravin Patil[2]

[1]*Mechanical Engineering Department, SCOE, Kharghar, Navi Mumbai, India*
[2]*Mechanical Engineering Department, VJTI, Matunga, Mumbai, India*

Abstract

Fuzzy sets help in dealing with the vagueness in data. An m-polar fuzzy set extends the set where a pole describes the relevance to the related property that controls the system. Since 2014, the m-polar fuzzy set is being used as a decision-making tool, and day by day, its usefulness is increasing. The challenge arises in using criteria weights while implementing m-polar fuzzy set hybrid methods. Researchers have either considered criteria weights directly or assumed their values. Thus, there is a need to obtain criteria weight while implementing various m-polar fuzzy set hybrid methodologies. For example, one can implement m-polar fuzzy set hybrid methods with AHP or Shannon's entropy weight calculation approach. In this chapter, Shannon's entropy weight calculation integrated m-polar fuzzy ELECTRE-I methodology is developed for solving a single-pole (single-valued) MCDM problem. Two industrial selection problems, i.e., selection of flexible manufacturing system and selection of cutting fluid, are considered for demonstration purposes. Work validation is done with a comparison of rankings of alternatives with earlier rankings. It is observed that Shannon's entropy weight calculation integrated m-polar fuzzy ELECTTRE-I concept is consistent with the original research done on the case study. Moreover, it will help deliver optimum results due to consideration of criteria weights during calculations.

***Keywords*:** MCDM, ELECTRE-I, m-polar, FMS, Shannon's entropy

**Corresponding author*: jagtap.aero@gmail.com

Sachi Nandan Mohanty, Prasenjit Chatterjee and Bui Thanh Hung (eds.) Fuzzy Computing in Data Science: Applications and Challenges, (113–146) © 2023 Scrivener Publishing LLC

7.1 Introduction

A bipolar fuzzy set is used to represent the positive and negative sides of the behavior of the system. Thus, a bipolar fuzzy set is a two-valued logic from two different agents (n=2). In real-time problems, the information can be collected from more than two agents ($n \geq 2$). From this notion, bipolar fuzzy set theory is further extended as m-polar fuzzy set theory. According to Chen [1], in the m-polar fuzzy set concept, the membership value of the element belongs to $[0, 1]^m$, where m represents various properties of that element.

It is revealed from the literature that the m-polar fuzzy set algebra helps solve multi-criteria group decision-making (MCGDM) problems consisting of incomplete data. One of the problems while using this approach is that the input data is considered in terms of the percentage of the actual performance. Another problem is that the performance of an alternative concerning one criterion may sometimes exceed that of the best option. To overcome all such limitations, researchers have developed hybrid methodologies. Different multi-criteria decision-making (MCDM) techniques are combined with m-polar fuzzy sets to evaluate MCGDM problems like selecting a location for the plant, selecting a suitable site for the airport, etc.

New MCDM techniques come with new algorithms and different approaches towards problem-solving. Therefore, the motivation for this research is to see the scope of the m-polar fuzzy ELECTRE-I method developed recently in 2019 [2]. Furthermore, the mechanical industry uses various MCDM techniques to solve industrial selection problems. Therefore, this chapter, exploring the scope of m-polar fuzzy ELECTRE-I to the mechanical industry selection problem.

7.1.1 Objectives

- To solve the problem of cutting fluid selection with the m-polar fuzzy ELECTRE-I method.
- To solve the problem of flexible manufacturing system with m-polar fuzzy ELECTRE-I method.
- To integrate Shannon's entropy weight calculation with m-polar fuzzy ELECTRE-I.
- To compare results obtained from m-polar fuzzy ELECTRE-I with techniques implemented by previous researchers.
- To obtain the outranking relationship between different available alternatives with the help of a directed graph.

7.2 Implementation of m-Polar Fuzzy ELECTRE-I Integrated Shannon's Entropy Weight Calculations

In this section, the detailed research methodology is explained step by step. It is a combination of the m-polar fuzzy ELECTRE-I method and Shannon's entropy weight calculation method. Various stages are described in equations, conditions, and matrices from the decision matrix to the final state of rank evaluation.

7.2.1 The m-Polar Fuzzy ELECTRE-I Integrated Shannon's Entropy Weight Calculation Method

Method is explained step by step as below:

1) Let $A = \{a_1, a_2, a_3 .... a_n\}$ set of options (Alternatives) available with $B = \{b_1, b_2, b_3, ... b_n\}$ set of criteria.
2) The Decision Matrix can be represented as the value of alternative to criteria with the help of

$$C = (c_{ij}) = \{c_{ij}{}^1, c_{ij}{}^2, c_{ij}{}^3 .... c_{ij}{}^m\}$$

3) Normalized Decision Matrix:

$$N = \frac{c_{ij}}{\sqrt{\sum_{i=1}^{m} c_{ij}^2}} \tag{7.1}$$

4) We are implementing Shannon's entropy weight calculation method for finding weights of Criteria.

$$e_j = -\frac{\sum_{i=1}^{m} f_{ij}\ lnf_{ij}}{ln\ m} \quad i = 1,2,3.....m \text{ and } j = 1,2,3....n \tag{7.2}$$

$$\text{Where } f_{ij} = \frac{r_{ij}}{\sum_{i=1}^{m} r_{ij}}$$

Entropy weights are calculated as:

$$w_j = \frac{1-e_j}{n-\sum_{i=1}^{m} e_j} \quad \text{where } \sum_{j=1}^{n} w_j = 1 \tag{7.3}$$

5) Applying weight to the m-polar decision matrix, $F = (f_{ij})$ is formed.

$$F = (f_{ij}) = (f_{ij}^{1}, f_{ij}^{2}, f_{ij}^{3} \ldots\ldots. f_{ij}^{m}) \text{ here } f_{ij} = w_j\, c_{ij}. \tag{7.4}$$

6) Concordance sets for m-polar fuzzy given by

$$G_{pq} = \{1 \leq j \leq t : y_{pj} \geq y_{qj}, p \neq q; p,q = 1,2\ldots\ldots\ldots.n\}$$

$$\text{Here, } y_{ij} = f_{ij}^{1} + f_{ij}^{2} + f_{ij}^{3} \ldots\ldots. + f_{ij}^{m}, \tag{7.5}$$

7) Discordance sets for m-polar fuzzy given by

$$H_{pq} = \{1 \leq j \leq t : y_{pj} \leq y_{qj}, p \neq q; p,q = 1,2\ldots\ldots\ldots.n\}$$

$$\text{Here, } y_{ij} = f_{ij}^{1} + f_{ij}^{2} + f_{ij}^{3} \ldots\ldots. + f_{ij}^{m}, \tag{7.6}$$

8) Concordance value of indices is formed as

$$g_{pq} = \sum\nolimits_{j \in g_{pq}} w_j, \text{ for all } p,q. \tag{7.7}$$

9) Concordance matrix G for the m-polar fuzzy set can be given as:

$$G = \begin{pmatrix} - & g_{12} & g_{13} & g_{1n} \\ g_{21} & - & \cdot\cdot & g_{2n} \\ \vdots & \vdots & \vdots & \vdots \\ g_{n1} & g_{n2} & \cdots & - \end{pmatrix}$$

10) Discordance indices value is formed as:

$$hpq = \frac{max_{j \in H_{pq}} \sqrt{\frac{1}{m}\left[\left(f^{1}_{pj} - f^{1}_{qj}\right)^2 + \left(f^{2}_{pj} - f^{2}_{qj}\right)^2 + \ldots + \left(f^{m}_{pj} - f^{m}_{qj}\right)^2\right]}}{max_{j} \sqrt{\frac{1}{m}\left[\left(f^{1}_{pj} - f^{1}_{qj}\right)^2 + \left(f^{2}_{pj} - f^{2}_{qj}\right)^2 + \ldots + \left(f^{m}_{pj} - f^{m}_{qj}\right)^2\right]}} \text{ for all } p,q \tag{7.8}$$

11) Discordance matrix H for the m-polar fuzzy set can be given as:

$$H = \begin{pmatrix} - & h_{12} & h_{13} & l_{1n} \\ h_{21} & - & .. & h_{2n} \\ : & : & : & : \\ h_{n1} & h_{n2} & \cdots & - \end{pmatrix}$$

12) Levels of Concordance and Discordance can be formulated as

$$\bar{g} = \frac{1}{n(n-1)} \sum\nolimits^{n}_{\substack{p=1 \\ p \neq q}} \sum\nolimits^{n}_{\substack{q=1 \\ q \neq p}} g_{pq} \tag{7.9}$$

$$\bar{h} = \frac{1}{n(n-1)} \sum\nolimits^{n}_{\substack{p=1 \\ p \neq q}} \sum\nolimits^{n}_{\substack{q=1 \\ q \neq p}} h_{pq} \tag{7.10}$$

13) From Concordance and Discordance levels, we can form Concordance dominance matrix and Discordance dominance matrix, which can be given as

$$S = \begin{pmatrix} - & s_{12} & s_{13} & s_{1n} \\ s_{21} & - & .. & s_{2n} \\ : & : & : & : \\ s_{n1} & s_{n2} & \cdots & - \end{pmatrix}$$

Here,

$$S_{pq} = \begin{cases} 1, \& g_{pq} \geq \bar{g} \\ 0, \& g_{pq} \geq \bar{g} \end{cases} \tag{7.11}$$

$$T = \begin{pmatrix} - & t_{12} & t_{13} & t_{1n} \\ t_{21} & - & .. & t_{2n} \\ : & : & : & : \\ t_{n1} & t_{n2} & \cdots & - \end{pmatrix}$$

Here,

$$t_{pq} = \begin{cases} 1, h_{pq} \geq \bar{h} \\ 0, h_{pq} \leq \bar{h} \end{cases} \tag{7.12}$$

14) To obtain aggregate dominance matrix V, perform point to point multiplication of K and U matrix values

$$U = \begin{pmatrix} - & u_{12} & u_{13} & u_{1n} \\ u_{21} & - & .. & u_{2n} \\ \vdots & \vdots & \vdots & \vdots \\ u_{n1} & u_{n2} & \cdots & - \end{pmatrix}$$

Above 14 steps is to be followed for solving the selection problem with the m-polar fuzzy ELECTRE-I method. Matrices S, T and U are used for outranking relations between alternatives.

7.3 Application to Industrial Problems

The research methodology was implemented through two industrial applications from literature and compared with results obtained from various MCDM techniques. In the following section, m-polar fuzzy ELECTRE-I integrated with Shannon's weight calculation implemented for cutting fluid selection and flexible manufacturing system selection problems.

7.3.1 Cutting Fluid Selection Problem

Several researchers worked on oil selection problems based on the properties of oils. They have implemented various methods for oil selection, as mentioned in Table 7.1. We have selected an issue from [3]; from the literature available, the previous author's had not used the ELECTRE-I method to evaluate rank for suitable cutting fluid. The specialty of the ELECTRE-I process is that we can find outranking relations between alternatives. The m-polar ELECTRE-I algorithm is helpful for MCDM and MCGDM.

Table 7.1 Methods applied for cutting fluid selection.

Sr. no.	Methods for cutting fluid selection	Application	References
1	digraph and matrix approach	Selection of cutting fluid for machining operation	[4]
2	two-grade fuzzy synthetic decision making system based on analytic hierarchy process (AHP)	Performance of grinding fluids	[5]
3	multi-objective decision making model	Cutting fluid selection	[6]
4	preference ranking organization method for enrichment evaluations (PROMETHEE)	Considering crisp and fuzzy criteria values for Choosing cutting fluid	[7]
5	software for optimization of the cutting fluid selection	Considering human safety and environmental hazards	[8]
6	combined AHP and Vlse Kriterijumska Optimizacija Kompromisno Resenje (VIKOR)	Considering the reduced cost, environmental impact and quality for selection of best-cutting fluid	[9]
7	decision support system based on AHP	To choose optimal cutting fluid for minimum environmental impact	[10]
8	multi-objective optimization on the basis of simple ratio analysis (MOOSRA) method	Considering environmental impact, cost and surface quality	[11]
9	Chi-square statistic and matrix approach	Selection of cutting fluid for given application	[12]
10	simple additive weighting (SAW) and weighted product (WPM) techniques	Ranking vegetable-based cutting fluid	[13]
11	quality function deployment (QFD)-based model	Selection of cutting fluids for specific machining requirements	[14]

In this problem from literature, a decision related to suitable cutting fluid for cutting spur gear from alloy steel as a material is evaluated with the m-polar fuzzy ELECTRE-I method. In the hobbing process, increase in the temperature adversely affects the surface finish and dimensional accuracy of spur gear. In several situations, fatty esters improve the coating ability to cut fluid. Cutting fluids with chlorine additives improves tool life at low speed. Cutting fluid with sulphur additives are used for machining high alloy steels. The viscosity of cutting fluid plays a vital role in protecting the environment, as it settles out suspensions of waste material from machining operations. For the current study, seven different cutting fluids are examined based on five properties. Viscosity at 40°C in cst, sulphur wt%, chlorine wt%, fatty ester wt%, and disposability in relative scale are used as the five criteria for selecting suitable cutting fluid. Table 7.2 shows the decision matrix for seven alternatives with five criteria of selection.

To provide suitable input, the m-polar fuzzy ELECTRE-I algorithm converting decision matrix in normalized form is shown in Table 7.3 with the help of equation 7.1.

Shannon's entropy weight calculations are used, and following weights of criteria obtained using equations 7.2 and 7.3, Table 7.4 shows the criteria used for selecting cutting fluid.

Table 7.2 Decision matrix for neat oil selection problem [3].

	Criteria				
	Viscosity at	**Sulphur wt%**	**Chlorine wt%**	**Fatty ester wt%**	**Disposability**
Alternatives	40°C (C_1)	(C_2)	(C_3)	(C_4)	(C_5)
A	12.5	2.8	1.8	0.5	4
B	31.6	3.5	1	3	5
C	13.5	2.3	1.1	5	5
D	24.1	2.5	2.7	5	4
E	45.1	2.4	2.6	5	3
F	38.5	2.2	2.4	4	3
G	21.6	1.1	3.5	7	4

Table 7.3 Normalized decision matrix for the neat oil selection problem.

Alternatives	Criteria				
	C_1	C_2	C_3	C_4	C_5
A	0.1627	0.4248	0.2939	0.0409	0.3714
B	0.4114	0.5310	0.1633	0.2456	0.4642
C	0.1757	0.3490	0.1796	0.4093	0.4642
D	0.3137	0.3793	0.4408	0.4093	0.3714
E	0.5871	0.3641	0.4245	0.4093	0.2785
F	0.5012	0.3338	0.3919	0.3274	0.2785
G	0.2812	0.1669	0.5715	0.5730	0.3714

Table 7.4 Weight calculated by Shannon's entropy method.

Criteria	C_1	C_2	C_3	C_4	C_5
Weights	0.0919	0.1621	0.2738	0.2748	0.1973

We can obtain a weight multiplied matrix with the help of equation 7.4. Table 7.5 shows the weight multiplied matrix.

Table 7.5 Weight multiplied matrix for cutting fluid selection problem.

Alternative	Criteria				
	C_1	C_2	C_3	C_4	C_5
A	0.0150	0.0689	0.0805	0.01123	0.0733
B	0.0378	0.0861	0.0447	0.0675	0.0916
C	0.0161	0.0566	0.0492	0.1124	0.0916
D	0.0288	0.0615	0.1206	0.1124	0.0733
E	0.0540	0.0590	0.1162	0.1124	0.0550
F	0.0460	0.0542	0.1073	0.0900	0.0550
G	0.0258	0.0271	0.1565	0.1575	0.0733

7.3.2 Results Obtained From m-Polar Fuzzy ELECTRE-I for Cutting Fluid Selection Problem

Weight multiplied matrix obtained from decision matrix and Shannon's entropy weight calculations is used as an input to the m-polar fuzzy ELECTRE-I method. Table 7.6 shows the concordance set obtained from the condition mentioned in step six of the research methodology.

Table 7.7 shows the discordance set obtained from the condition mentioned in step seven of the research methodology

Table 7.6 Concordance set.

j	1	2	3	4	5	6	7
G_{1j}	-	{3}	{2,3}	{2,5}	{2,5}	{2,5}	{2,5}
G_{2j}	{1,2,4,5}	-	{1,2,5}	{1,2,5}	{2,5}	{2,5}	{1,2,5}
G_{3j}	{1,4,5}	{3,4,5}	-	{4,5}	{4,5}	{2,4,5}	{2,5}
G_{4j}	{1,3,4,5}	{3,4}	{1,2,3,4}	-	{2,3,4,5}	{2,3,4,5}	{1,2,5}
G_{5j}	{1,3,4}	{1,3,4}	{1,2,3,4}	{1,4}	-	{1,2,3,4,5}	{1,2}
G_{6j}	{1,3,4}	{1,3,4}	{1,3}	{1}	{5}	-	{1,2}
G_{7j}	{1,3,4,5}	{3,4}	{1,3,4}	{3,4,5}	{3,4,5}	{3,4,5}	-

Table 7.7 Discordance set.

j	1	2	3	4	5	6	7
H_{1j}	-	{1,2,4,5}	{1,4,5}	{1,3,4}	{1,3,4}	{1,3,4}	{1,3,4}
H_{2j}	{3}	-	{3,4}	{3,4}	{1,3,4}	{1,3,4}	{3,4}
H_{3j}	{2,3}	{1,2}	-	{1,2,3}	{1,2,3}	{1,3}	{1,3,4}
H_{4j}	{2}	{1,2,5}	{5}	-	{1}	{1}	{3,4}
H_{5j}	{2,5}	{2,5}	{5}	{2,3,5}	-	{}	{3,4,5}
H_{6j}	{2,5}	{2,5}	{2,4,5}	{2,3,4,5}	{1,2,3,4}	-	{3,4,5}
H_{7j}	{2}	{1,2,5}	{2,5}	{1,2}	{1,2}	{1,2}	-

Matrix G, shown below, is obtained from the concordance indices with equation seven from research methodology. Similarly, matrix H is obtained from discordance indices with equation eight from research methodology.

$$G = \begin{bmatrix} - & 0.2738 & 0.4359 & 0.3594 & 0.3594 & 0.3594 & 0.3594 \\ 0.7261 & - & 0.4513 & 0.4513 & 0.3594 & 0.3594 & 0.4513 \\ 0.564 & 0.7459 & - & 0.4721 & 0.4721 & 0.6342 & 0.3594 \\ 0.8378 & 0.5486 & 0.8026 & - & 0.908 & 0.908 & 0.4513 \\ 0.6405 & 0.6405 & 0.8026 & 0.3667 & - & 0.9999 & 0.254 \\ 0.6405 & 0.6405 & 0.3657 & 0.0919 & 0.1973 & - & 0.254 \\ 0.8378 & 0.5486 & 0.6405 & 0.7459 & 0.7459 & 0.7459 & - \end{bmatrix}$$

$$H = \begin{bmatrix} - & 1 & 1 & 1 & 1 & 1 & 1 \\ 0.6356 & - & 1 & 1 & 1 & 1 & 1 \\ 0.3091 & 0.6558 & - & 1 & 1 & 1 & 1 \\ 0.0728 & 0.3236 & 0.2560 & - & 1 & 0.7656 & 1 \\ 0.1810 & 0.5123 & 0.5464 & 0.7295 & - & 0 & 1 \\ 0.2328 & 0.5853 & 0.6303 & 1 & 1 & - & 1 \\ 0.2859 & 0.5280 & 0.2750 & 0.7653 & 0.7105 & 0.4008 & - \end{bmatrix}$$

Concordance level and discordance level are obtained from equation nine and equation ten from research methodology. Their values are $\bar{g} = 0.543067$ and $\bar{h} = 0.723868$ respectively. Concordance level and discordance levels are used to obtain matrix S and matrix T, respectively. When the value of concordance indices is less than concordance level, then the value of the element for matrix S will be zero; otherwise, it's one. Similarly, for matrix T, when the value of discordance indices is more than discordance level, then the value of the element in matrix T is zero; otherwise, it's one.

$$S = \begin{bmatrix} 0 & 0 & 0 & 0 & 0 & 0 & 0 \\ 1 & 0 & 0 & 0 & 0 & 0 & 0 \\ 1 & 1 & 0 & 0 & 0 & 1 & 0 \\ 1 & 1 & 1 & 0 & 1 & 1 & 0 \\ 1 & 1 & 1 & 0 & 0 & 1 & 0 \\ 1 & 1 & 0 & 0 & 0 & 0 & 0 \\ 1 & 1 & 1 & 1 & 1 & 1 & 0 \end{bmatrix}$$

$$T = \begin{bmatrix} 0 & 0 & 0 & 0 & 0 & 0 & 0 \\ 1 & 0 & 0 & 0 & 0 & 0 & 0 \\ 1 & 1 & 0 & 0 & 0 & 0 & 0 \\ 1 & 1 & 1 & 0 & 0 & 0 & 0 \\ 1 & 1 & 1 & 0 & 0 & 1 & 0 \\ 1 & 1 & 1 & 0 & 0 & 0 & 0 \\ 1 & 1 & 1 & 0 & 1 & 1 & 0 \end{bmatrix}$$

Matrix U is the point to point multiplicative form of matrix S and matrix T. The combination of values obtained from matrix S, T, and U are used to find outranking relationships between variables, as shown in Table 7.9.

$$U = \begin{bmatrix} 0 & 0 & 0 & 0 & 0 & 0 & 0 \\ 1 & 0 & 0 & 0 & 0 & 0 & 0 \\ 1 & 1 & 0 & 0 & 0 & 0 & 0 \\ 1 & 1 & 1 & 0 & 0 & 0 & 0 \\ 1 & 1 & 1 & 0 & 0 & 1 & 0 \\ 1 & 1 & 0 & 0 & 0 & 0 & 0 \\ 1 & 1 & 1 & 0 & 1 & 1 & 0 \end{bmatrix}$$

G is the best alternative for the current problem as it has advantages over the other available options. Table 7.8 below shows a comparison of different methods for rank performance of cutting fluid selection problem. We can observe clearly that the modified similarity method and TOPSIS method have similar rankings. At the same time, the m-polar fuzzy ELECTRE-I algorithm approach gives the same order except for the second and third rank for alternatives. It shows that the single-valued input to the m-polar fuzzy ELECTRE-I integrated with Shannon's entropy weight calculations gives similar results as the modified similarity method and TOPSIS method. The advantage of the m-polar fuzzy ELECTRE-I method over the other methods is that it ranks relationships between alternatives. With the help of outranking relations, we can plot a directed graph, as shown in Figure 7.1 below. The arrow in the directed graph indicates the dominance of one method over the other methods. From the directed graph, we can conclude that alternative G is the best alternative for cutting fluid selection. It is validated from the comparison with the results obtained from different methods.

Table 7.8 Comparison between modified similarity and TOPSIS methods for neat oil selection.

Problem alternatives	Modified similarity method	TOPSIS	m-polar fuzzy ELECTRE-I
A	7	7	7
B	6	6	6
C	5	5	5
D	2	2	3
E	3	3	2
F	4	4	4
G	1	1	1

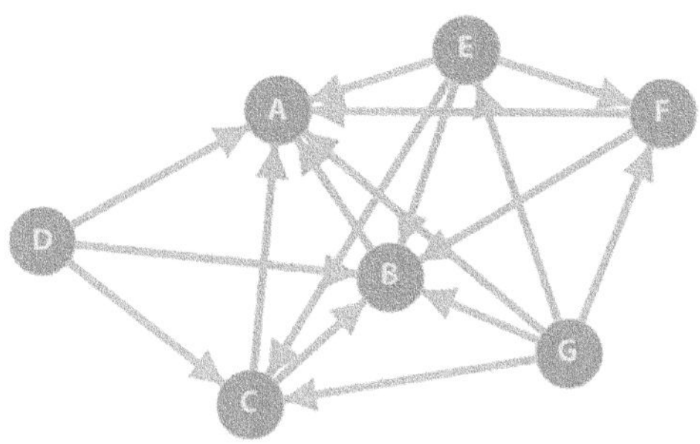

Figure 7.1 Directed graph for cutting fluid selection.

Table 7.9 below shows outranking relationships between alternatives; we can read this relationship as when the s, t, and u are unity, then there will be a comparison between other options; otherwise, they are incomparable. Alternative G shows maximum relationships with other alternatives; therefore, it is the best alternative.

7.3.3 FMS Selection Problem

Flexible manufacturing systems (FMS) are capable of processing various part styles at different workstations. As a result, FMS can produce a range of products in batch size. Batch size may vary from zero to thousand. Several researchers used different MCDM techniques for the selection of

Table 7.9 Outranking relationship between alternatives.

Comparison of cutting fluids	G_{pq}	H_{pq}	g_{pq}	h_{pq}	s	t	u	Ranking
(A,B)	{3}	{1,2,4,5}	0.2738	1	0	0	0	Incomparable
(A,C)	{2,3}	{1,4,5}	0.4359	1	0	0	0	Incomparable
(A,D)	{2,5}	{1,3,4}	0.3594	1	0	0	0	Incomparable
(A,E)	{2,5}	{1,3,4}	0.3594	1	0	0	0	Incomparable
(A,F)	{2,5}	{1,3,4}	0.3594	1	0	0	0	Incomparable
(A,G)	{2,5}	{1,3,4}	0.3594	1	0	0	0	Incomparable
(B,A)	{1,2,4,5}	{3}	0.7261	0.6356	1	1	1	B -- A
(B,C)	{1,2,5}	{3,4}	0.4513	1	0	0	0	Incomparable
(B,D)	{1,2,5}	{3,4}	0.4513	1	0	0	0	Incomparable
(B,E)	{2,5}	{1,3,4}	0.3594	1	0	0	0	Incomparable
(B,F)	{2,5}	{1,3,4}	0.3594	1	0	0	0	Incomparable
(B,G)	{1,2,5}	{3,4}	0.4513	1	0	0	0	Incomparable

(Continued)

Table 7.9 Outranking relationship between alternatives. (*Continued*)

Comparison of cutting fluids	G_{pq}	H_{pq}	g_{pq}	h_{pq}	s	t	u	Ranking
(C,A)	{1,4,5}	{2,3}	0.564	0.3091	1	1	1	C -- A
(C,B)	{3,4,5}	{1,2}	0.7459	0.6558	1	1	1	C -- B
(C,D)	{4,5}	{1,2,3}	0.4721	1	0	0	0	Incomparable
(C,E)	{4,5}	{1,2,3}	0.4721	1	0	0	0	Incomparable
(C,F)	{2,4,5}	{1,3}	0.6342	1	1	0	0	Incomparable
(C,G)	{2,5}	{1,3,4}	0.3594	1	0	0	0	Incomparable
(D,A)	{1,3,4,5}	{2}	0.8378	0.0728	1	1	1	D -- A
(D,B)	{3,4}	{1,2,5}	0.5486	0.3236	1	1	1	D -- B
(D,C)	{1,2,3,4}	{5}	0.8026	0.2560	1	1	1	D -- C
(D,E)	{2,3,4,5}	{1}	0.908	1	1	0	0	Incomparable
(D,F)	{2,3,4,5}	{1}	0.908	0.7656	1	0	0	Incomparable
(D,G)	{1,2,5}	{3,4}	0.4513	1	0	0	0	Incomparable
(E,A)	{1,3,4}	{2,5}	0.6405	0.1810	1	1	1	E -- A

(*Continued*)

Table 7.9 Outranking relationship between alternatives. (*Continued*)

Comparison of cutting fluids	G_{pq}	H_{pq}	g_{pq}	h_{pq}	s	t	u	Ranking
(E,B)	{1,3,4}	{2,5}	0.6405	0.5123	1	1	1	E -- B
(E,C)	{1,2,3,4}	{5}	0.8026	0.5464	1	1	1	E -- C
(E,D)	{1,4}	{2,3,5}	0.3667	0.7295	0	0	0	Incomparable
(E,F)	{1,2,3,4,5}	{}	0.9999	0	1	1	1	E -- F
(E,G)	{1,2}	{3,4,5}	0.254	1	0	0	0	Incomparable
(F,A)	{1,3,4}	{2,5}	0.6405	0.2328	1	1	1	F -- A
(F,B)	{1,3,4}	{2,5}	0.6405	0.5853	1	1	1	F -- B
(F,C)	{1,3}	{2,4,5}	0.3657	0.6303	0	1	0	Incomparable
(F,D)	{1}	{2,3,4,5}	0.0919	1	0	0	0	Incomparable
(F,E)	{5}	{1,2,3,4}	0.1973	1	0	0	0	Incomparable
(F,G)	{1,2}	{3,4,5}	0.254	1	0	0	0	Incomparable
(G,A)	{1,3,4,5}	{2}	0.8378	0.2859	1	1	1	G -- A
(G,B)	{3,4}	{1,2,5}	0.5486	0.5280	1	1	1	G -- B

(*Continued*)

Table 7.9 Outranking relationship between alternatives. (*Continued*)

Comparison of cutting fluids	$\mathbf{G_{pq}}$	$\mathbf{H_{pq}}$	$\mathbf{g_{pq}}$	$\mathbf{h_{pq}}$	**s**	**t**	**u**	**Ranking**
(G,C)	{1,3,4}	{2,5}	0.6405	0.2750	1	1	1	G -- C
(G,D)	{3,4,5}	{1,2}	0.7459	0.7653	1	0	0	Incomparable
(G,E)	{3,4,5}	{1,2}	0.7459	0.7105	1	1	1	G -- E
(G,F)	{3,4,5}	{1,2}	0.7459	0.4008	1	1	1	G -- F

Table 7.10 Methods for FMS selection.

Sr. no.	Method for FMS selection	Application	References
1	Analytical Hierarchy Process	Selection of flexible manufacturing System (FMS)	[16–18]
2	Digraph and matrix approach		[19, 20]
3	MACBETH		[21]
5	Data envelopment analysis		[22–26]
6	Technique for order preference by similarity to ideal solution (TOPSIS)		[27, 28]
7	Euclidean distance-based integrated approach		[29]

FMS, as shown in Table 7.10. In the current study, the FMS selection problem is selected from [27], the m-polar fuzzy ELECTRE-I algorithm is used with single-valued input.

FMS selection problems were selected from Karsak [27], eight alternatives and eight criteria for FMS selection. For optimum investment in FMS, Karsak [27] fixed capital and operating cost (COS), required floor space (RFS), work in progress (WIP), product flexibility (PF), volume flexibility (VF), expansion flexibility (EF), lead time reduction (LTR), and quality improvement (QI) as variables. Product flexibility provides starts for new products quickly and economically. Volume flexibility gives profitability at varying production levels. Finally, expansion flexibility adds capacity and capability to the existing system. Out of eight criteria, three are quantitative, and five are taken in the range of [0,1]. Table 7.11 shows the decision matrix for the FMS selection problem.

For providing suitable input to the m-polar fuzzy ELECTRE-I algorithm converting decision matrix in normalized form as shown in Table 7.12 with the help of equation 7.1.

Shannon's entropy weight calculations are used, and following weights of criteria obtained using equation two and equation three, Table 7.13 shows weights of criteria used for selection of FMS.

We can obtain a weight multiplied matrix with the help of equation four; Table 7.14 shows the weight multiplied matrix.

7.3.4 Results Obtained From m-Polar Fuzzy ELECTRE-I for FMS Selection

Weight multiplied matrix obtained from decision matrix and Shannon's entropy weight calculations is used as an input to the m-polar fuzzy ELECTRE-I method. Table 7.15 shows the concordance set obtained from the condition mentioned in step six of the research methodology.

Table 7.11 Decision matrix for FMS selection problem [27].

FMS	Capital and operating costs ($ millions)	Required floor space (m^2)	Work in progress (units)	Product flexibility	Volume flexibility	Expansion flexibility	Lead time reduction	Quality improvement
A	3.8	630	42	0.7	0.5	0.7	0.5	0.2
B	3.1	620	37	0.7	0.7	0.5	0.9	0.5
C	5	425	32	0.9	0.7	0.7	0.7	0.7
D	6.4	500	54	0.7	0.5	0.5	0.7	0.9
E	3.6	600	44	0.2	0.7	0.7	0.2	0.7
F	6.7	780	59	0.9	0.9	0.5	0.5	0.5
G	3.4	740	37	0.5	0.2	0.2	0.7	0.9
H	3.7	550	36	0.5	0.9	0.9	0.5	0.5

Table 7.12 Normalized decision matrix for the FMS selection problem.

FMS	COS	RFS	WIP	PF	VF	EF	LTR	QI
A	0.8056	0.4225	0.6296	0.7143	0.4286	0.7143	0.4286	0
B	1	0.4507	0.8148	0.7143	0.7143	0.4286	1	0.4286
C	0.4722	1	1	1	0.7143	0.7143	0.7143	0.7143
D	0.0833	0.7887	0.1852	0.7143	0.4286	0.4286	0.7143	1
E	0.8611	0.5070	0.5556	0	0.7143	0.7143	0	0.7143
F	0	0	0	1	1	0.4286	0.4286	0.4286
G	0.9167	0.1127	0.8148	0.4286	0	0	0.7143	1
H	0.8333	0.6479	0.8519	0.4286	1	1	0.4286	0.4286

Table 7.13 Shannon's entropy weight for variables

Criteria	COS	RFS	WIP	PF	VF	EF	LTR	QI
Weight	0.1297	0.1420	0.1400	0.1173	0.1173	0.1194	0.1194	0.1150

Table 7.14 Weight multiplied matrix for FMS selection problem.

FMS	COS	RFS	WIP	PF	VF	EF	LTR	QI
A	0.0138	0.0185	0.0172	0.0161	0.0115	0.0178	0.0127	0.0047
B	0.0113	0.0182	0.0152	0.0161	0.0161	0.0127	0.0229	0.0117
C	0.0182	0.0125	0.0131	0.0207	0.0161	0.0178	0.0178	0.0164
D	0.0233	0.0147	0.0222	0.0161	0.0115	0.0127	0.0178	0.0211
E	0.0131	0.0176	0.0181	0.0046	0.0161	0.0178	0.0051	0.0164
F	0.0243	0.0229	0.0242	0.0207	0.0207	0.0127	0.0127	0.0117
G	0.0124	0.0217	0.0152	0.0115	0.0046	0.0051	0.0178	0.0211
H	0.0134	0.0161	0.0148	0.0115	0.0207	0.0229	0.0127	0.0117

Table 7.15 Concordance set.

	1	2	3	4	5	6	7	8
G_{1j}	-	{4,6}	{1,6}	{1,3,4,5,6}	{3,4,6,7}	{1,2,3,6,7}	{2,4,5,6}	{4,7}
G_{2j}	{1,2,3,4,5,7,8}	-	{1,5,7}	{1,3,4,5,6,7}	{1,3,4,5,7}	{1,2,3,6,7,8}	{1,2,3,4,5,6,7}	{1,4,7,8}
G_{3j}	{2,3,4,5,6,7,8}	{2,3,4,5,6,8}	-	{1,2,3,4,5,6,7}	{2,3,4,5,6,7,8}	{1,2,3,4,6,7,8}	{2,3,4,5,6,7}	{2,3,4,7,8}
G_{4j}	{2,4,5,7,8}	{2,4,6,8}	{7,8}	-	{2,4,7,8}	{1,2,3,6,7,8}	{2,4,5,6,7,8}	{2,4,7,8}
G_{5j}	{1,2,5,6,8}	{2,5,6,8}	{1,5,6,8}	{1,3,5,6}	-	{1,2,3,6,8}	{2,5,6}	{1,8}
G_{6j}	{4,5,7,8}	{4,5,6,8}	{4,5}	{4,5,6}	{4,5,7}	-	{4,5,6}	{4,5,7,8}
G_{7j}	{1,3,7,8}	{3,8}	{1,7,8}	{1,3,7,8}	{1,3,4,7,8}	{1,2,3,7,8}	-	{1,4,7,8}
G_{8j}	{1,2,3,5,6,7,8}	{2,3,5,6,8}	{1,5,6}	{1,3,5,6}	{2,3,4,5,6,7}	{1,2,3,5,6,7,8}	{2,3,4,5,6}	-

Table 7.16 Discordance set.

	1	**2**	**3**	**4**	**5**	**6**	**7**	**8**
$\mathbf{H}_{1j}$	-	{1,2,3,5,7,8}	{2,3,4,5,7,8}	{2,7,8}	{1,2,5,8}	{4,5,8}	{1,3,7,8}	{1,2,3,5,6,8}
$\mathbf{H}_{2j}$	{6}	-	{2,3,4,6,8}	{2,8}	{2,6,8}	{4,5}	{8}	{2,3,5,6}
$\mathbf{H}_{3j}$	{1}	{1,7}	-	{8}	{1}	{5}	{1,8}	{1,5,6}
$\mathbf{H}_{4j}$	{1,3,6}	{1,3,5,7}	{1,2,3,4,5,6}	-	{1,3,5,6}	{4,5}	{1,3}	{1,3,5,6}
$\mathbf{H}_{5j}$	{3,4,7}	{1,3,4,7}	{2,3,4,7}	{2,4,7,8}	-	{4,5,7}	{1,3,4,7,8}	{2,3,4,5,6,7}
$\mathbf{H}_{6j}$	{1,2,3,6}	{1,2,3,7}	{1,2,3,6,7,8}	{1,2,3,7,8}	{1,2,3,6,8}	-	{1,2,3,7,8}	{1,2,3,6}
$\mathbf{H}_{7j}$	{2,4,5,6}	{1,2,4,5,6,7}	{2,3,4,5,6}	{2,4,5,6}	{2,5,6}	{4,5,6}	-	{2,3,5,6}
$\mathbf{H}_{8j}$	{4}	{1,4,7}	{2,3,4,7,8}	{2,4,7,8}	{1,8}	{4}	{1,7,8}	-

Table 7.16 shows the discordance set obtained from the condition mentioned in step seven of the research methodology.

Matrix G, shown below, is obtained from the concordance indices with equation seven from research methodology. Similarly, matrix H is obtained from discordance indices with equation eight from research methodology.

$$G = \begin{bmatrix} - & 0.2367 & 0.2491 & 0.6237 & 0.4961 & 0.6505 & 0.496 & 0.2367 \\ 0.8807 & - & 0.3664 & 0.7431 & 0.6237 & 0.7655 & 0.8851 & 0.4814 \\ 0.8704 & 0.751 & - & 0.88851 & 0.8704 & 0.8828 & 0.7554 & 0.6337 \\ 0.611 & 0.4937 & 0.2344 & - & 0.4937 & 0.7655 & 0.7304 & 0.4937 \\ 0.6234 & 0.4937 & 0.4814 & 0.5064 & - & 0.6461 & 0.3787 & 0.2447 \\ 0.469 & 0.469 & 0.2346 & 0.354 & 0.354 & - & 0.354 & 0.469 \\ 0.5041 & 0.255 & 0.3641 & 0.5041 & 0.6241 & 0.6461 & - & 0.4814 \\ 0.8828 & 0.6337 & 0.3664 & 0.5064 & 0.7554 & 0.8828 & 0.636 & - \end{bmatrix}$$

$$H = \begin{bmatrix} - & 1 & 1 & 1 & 0.9803 & 0.6414 & 1 & 1 \\ 0.5 & - & 1 & 0.5526 & 0.2857 & 0.2583 & 0.7842 & 1 \\ 0.5264 & 0.8776 & - & 0.2880 & 0.4300 & 0.2360 & 0.4575 & 0.6987 \\ 0.8146 & 1 & 1 & - & 1 & 0.5984 & 1 & 1 \\ 1 & 1 & 1 & 0.8454 & - & 1 & 1 & 1 \\ 1 & 1 & 1 & 1 & 0.9521 & - & 1 & 1 \\ 0.7416 & 1 & 1 & 0.8880 & 1 & 0.9865 & - & 1 \\ 0.5 & 1 & 1 & 0.6755 & 0.6420 & 0.5619 & 0.5503 & - \end{bmatrix}$$

Concordance level and Discordance level are obtained from equation nine and equation ten from research methodology. Their values are $\bar{g} = 0.55935$ and $\bar{h} = 0.826323$ respectively. Concordance level and discordance levels are used to obtain matrix S and matrix T, respectively. When the value of concordance indices is less than concordance level, then the value of the element for matrix S will be 0; otherwise, it is 1. Similarly, for matrix T, when the value of discordance indices is more than discordance level, then the value of the element in matrix T is 0; otherwise, it is 1.

$$S = \begin{bmatrix} 0 & 0 & 0 & 1 & 0 & 1 & 0 & 0 \\ 1 & 0 & 0 & 1 & 1 & 1 & 1 & 0 \\ 1 & 1 & 0 & 1 & 1 & 1 & 1 & 1 \\ 1 & 0 & 0 & 0 & 0 & 1 & 1 & 0 \\ 1 & 0 & 0 & 0 & 0 & 1 & 0 & 0 \\ 0 & 0 & 0 & 0 & 0 & 0 & 0 & 0 \\ 0 & 0 & 0 & 0 & 1 & 1 & 0 & 0 \\ 1 & 1 & 0 & 0 & 1 & 1 & 1 & 0 \end{bmatrix}$$

$$T = \begin{bmatrix} 0 & 0 & 0 & 0 & 0 & 1 & 0 & 0 \\ 1 & 0 & 0 & 1 & 1 & 1 & 1 & 0 \\ 1 & 0 & 0 & 1 & 1 & 1 & 1 & 1 \\ 1 & 0 & 0 & 0 & 0 & 1 & 0 & 0 \\ 0 & 0 & 0 & 0 & 0 & 0 & 0 & 0 \\ 0 & 0 & 0 & 0 & 0 & 0 & 0 & 0 \\ 1 & 0 & 0 & 0 & 0 & 0 & 0 & 0 \\ 1 & 0 & 0 & 1 & 1 & 1 & 1 & 0 \end{bmatrix}$$

Matrix U is the point to point multiplicative form of matrix S and matrix T. The combination of values obtained from matrix S, T, and U are used to find outranking relationships between variables, as shown in Table 7.18.

$$U = \begin{bmatrix} 0 & 0 & 0 & 0 & 0 & 1 & 0 & 0 \\ 1 & 0 & 0 & 1 & 1 & 1 & 1 & 0 \\ 1 & 0 & 0 & 1 & 1 & 1 & 1 & 1 \\ 1 & 0 & 0 & 0 & 0 & 1 & 0 & 0 \\ 0 & 0 & 0 & 0 & 0 & 0 & 0 & 0 \\ 0 & 0 & 0 & 0 & 0 & 0 & 0 & 0 \\ 0 & 0 & 0 & 0 & 0 & 0 & 0 & 0 \\ 1 & 0 & 0 & 0 & 1 & 1 & 1 & 0 \end{bmatrix}$$

C is the best alternative for the current problem as it has advantages over the other available options. Table 7.17 below shows the comparison of different methods for rank performance of the FMS selection problem. We can observe clearly that the EVAMIX method and COPRAS method have other rankings. At the same time, the m-polar fuzzy ELECTRE-I algorithm approach gives the same order except for B, E, F, and H alternatives. It shows that the single-valued input to the m-polar fuzzy ELECTRE-I integrated with Shannon's entropy weight calculations gives similar results as the COPRAS method. The advantage of the m-polar fuzzy ELECTRE-I method over the other methods is that it is outranking relationships between alternatives. With the help of outranking relations, we can plot a directed graph, as shown in Figure 7.2 below. The arrow in the directed graph indicates the dominance of one method over the other methods. From the directed graph, we can conclude that alternative C is the best

Table 7.17 Comparison between EVAMIX, COPRAS, and m-polar fuzzy ELECTRE-I methods for FMS selection.

FMS methods	EVAMIX	COPRAS	m-polar fuzzy ELECTRE-I
1	8	5	5
2	3	3	2
3	1	1	1
4	5	4	4
5	4	7	6
6	6	6	7
7	7	8	8
8	2	2	3

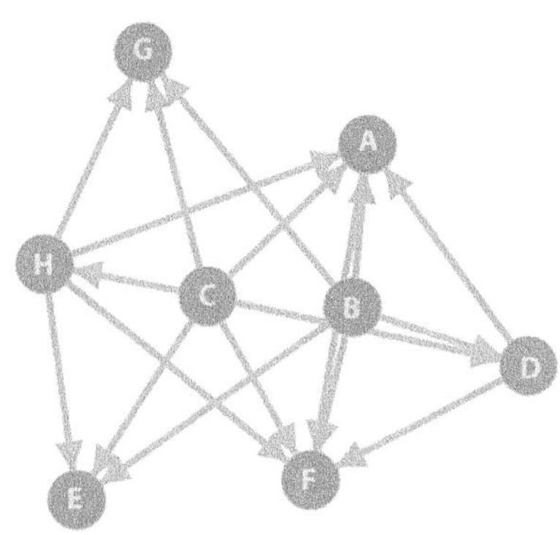

Figure 7.2 Directed graph for FMS selection.

alternative for FMS selection. It is validated from the comparison with the results obtained from different methods. Other industrial applications of m-polar fuzzy ELECTRE-I are explained in Jagtap and Karande [30].

Table 7.18 below shows outranking relationships between alternatives; we can read this relationship as when the s, t, and u are unity, then there will be a comparison between options; otherwise, they are incomparable. So, for example, alternative C is the best.

Table 7.18 Outranking relationship between alternatives.

Comparison of FMS	G_{pq}	H_{pq}	g_{pq}	h_{pq}	s	t	u	Ranking
(A,B)	{4,6}	{1,2,3,5,7,8}	0.2367	1	0	0	0	Incomparable
(A,C)	{1,6}	{2,3,4,5,7,8}	0.2491	1	0	0	0	Incomparable
(A,D)	{1,3,4,5,6}	{2,7,8}	0.6237	1	1	0	0	Incomparable
(A,E)	{3,4,6,7}	{1,2,5,8}	0.4961	0.9803	0	0	0	Incomparable
(A,F)	{1,2,3,6,7}	{4,5,8}	0.6505	0.6414	1	1	1	A -- F
(A,G)	{2,4,5,6}	{1,3,7,8}	0.496	1	0	0	0	Incomparable
(A,H)	{4,7}	{1,2,3,5,6,8}	0.2367	1	0	0	0	Incomparable
(B,A)	{1,2,3,4,5,7,8}	{6}	0.8807	0.5	1	1	1	B -- A
(B,C)	{1,5,7}	{2,3,4,6,8}	0.3664	1	0	0	0	Incomparable
(B,D)	{1,3,4,5,6,7}	{2,8}	0.7431	0.5526	1	1	1	B -- D
(B,E)	{1,3,4,5,7}	{2,6,8}	0.6237	0.2857	1	1	1	B -- E
(B,F)	{1,2,3,6,7,8}	{4,5}	0.7655	0.2583	1	1	1	B -- F
(B,G)	{1,2,3,4,5,6,7}	{8}	0.8851	0.7842	1	1	1	B -- G

(Continued)

Table 7.18 Outranking relationship between alternatives. (*Continued*)

Comparison of FMS	G_{pq}	H_{pq}	g_{pq}	h_{pq}	s	t	u	Ranking
(B,H)	{1,4,7,8}	{2,3,5,6}	0.4814	1	0	0	0	Incomparable
(C,A)	{2,3,4,5,6,7,8}	{1}	0.8704	0.5264	1	1	1	C -- A
(C,B)	{2,3,4,5,6,8}	{1,7}	0.751	0.8776	1	0	0	Incomparable
(C,D)	{1,2,3,4,5,6,7}	{8}	0.8851	0.2880	1	1	1	C -- D
(C,E)	{2,3,4,5,6,7,8}	{1}	0.8704	0.4300	1	1	1	C -- E
(C,F)	{1,2,3,4,6,7,8}	{5}	0.8828	0.2360	1	1	1	C -- F
(C,G)	{2,3,4,5,6,7}	{1,8}	0.7554	0.4575	1	1	1	C -- G
(C,H)	{2,3,4,7,8}	{1,5,6}	0.6337	0.6987	1	1	1	C -- H
(D,A)	{2,4,5,7,8}	{1,3,6}	0.611	0.8146	1	1	1	D -- A
(D,B)	{2,4,6,8}	{1,3,5,7}	0.4937	1	0	0	0	Incomparable
(D,C)	{7,8}	{1,2,3,4,5,6}	0.2344	1	0	0	0	Incomparable
(D,E)	{2,4,7,8}	{1,3,5,6}	0.4937	1	0	0	0	Incomparable
(D,F)	{1,2,3,6,7,8}	{4,5}	0.7655	0.5984	1	1	1	D -- F

(*Continued*)

Table 7.18 Outranking relationship between alternatives. (*Continued*)

Comparison of FMS	G_{pq}	H_{pq}	g_{pq}	h_{pq}	s	t	u	Ranking
(D,G)	{2,4,5,6,7,8}	{1,3}	0.7304	1	1	0	0	Incomparable
(D,H)	{2,4,7,8}	{1,3,5,6}	0.4937	1	0	0	0	Incomparable
(E,A)	{1,2,5,6,8}	{3,4,7}	0.6234	1	1	0	0	Incomparable
(E,B)	{2,5,6,8}	{1,3,4,7}	0.4937	1	0	0	0	Incomparable
(E,C)	{1,5,6,8}	{2,3,4,7}	0.4814	1	0	0	0	Incomparable
(E,D)	{1,3,5,6}	{2,4,7,8}	0.5064	0.8454	0	0	0	Incomparable
(E,F)	{1,2,3,6,8}	{4,5,7}	0.6461	1	1	0	0	Incomparable
(E,G)	{2,5,6}	{1,3,4,7,8}	0.3787	1	0	0	0	Incomparable
(E,H)	{1,8}	{2,3,4,5,6,7}	0.2447	1	0	0	0	Incomparable
(F,A)	{4,5,7,8}	{1,2,3,6}	0.469	1	0	0	0	Incomparable
(F,B)	{4,5,6,8}	{1,2,3,7}	0.469	1	0	0	0	Incomparable
(F,C)	{4,5}	{1,2,3,6,7,8}	0.2346	1	0	0	0	Incomparable
(F,D)	{4,5,6}	{1,2,3,7,8}	0.354	1	0	0	0	Incomparable

(*Continued*)

Table 7.18 Outranking relationship between alternatives. (*Continued*)

Comparison of FMS	G_{pq}	H_{pq}	g_{pq}	h_{pq}	s	t	u	Ranking
(F,E)	{4,5,7}	{1,2,3,6,8}	0.354	0.9521	0	0	0	Incomparable
(F,G)	{4,5,6}	{1,2,3,7,8}	0.354	1	0	0	0	Incomparable
(F,H)	{4,5,7,8}	{1,2,3,6}	0.469	1	0	0	0	Incomparable
(G,A)	{1,3,7,8}	{2,4,5,6}	0.5041	0.7416	0	1	0	Incomparable
(G,B)	{3,8}	{1,2,4,5,6,7}	0.255	1	0	0	0	Incomparable
(G,C)	{1,7,8}	{2,3,4,5,6}	0.3641	1	0	0	0	Incomparable
(G,D)	{1,3,7,8}	{2,4,5,6}	0.5041	0.8880	0	0	0	Incomparable
(G,E)	{1,3,4,7,8}	{2,5,6}	0.6241	1	1	0	0	Incomparable
(G,F)	{1,2,3,7,8}	{4,5,6}	0.6461	0.9865	1	0	0	Incomparable
(G,H)	{1,4,7,8}	{2,3,5,6}	0.4814	1	0	0	0	Incomparable
(H,A)	{1,2,3,5,6,7,8}	{4}	0.8828	0.5	1	1	1	H -- A
(H,B)	{2,3,5,6,8}	{1,4,7}	0.6337	1	1	0	0	Incomparable

(*Continued*)

Table 7.18 Outranking relationship between alternatives. (*Continued*)

Comparison of FMS	G_{pq}	H_{pq}	g_{pq}	h_{pq}	**s**	**t**	**u**	**Ranking**
(H,C)	{1,5,6}	{2,3,4,7,8}	0.3664	1	0	0	0	Incomparable
(H,D)	{1,3,5,6}	{2,4,7,8}	0.5064	0.6755	0	1	0	Incomparable
(H,E)	{2,3,4,5,6,7}	{1,8}	0.7554	0.6420	1	1	1	H -- E
(H,F)	{1,2,3,5,6,7,8}	{4}	0.8828	0.5619	1	1	1	H -- F
(H,G)	{2,3,4,5,6}	{1,7,8}	0.636	0.5503	1	1	1	H -- G

7.4 Conclusions

In this chapter, the m-polar fuzzy ELECTRE-I integrated Shannon's entropy method is implemented to cut fluid selection and FMS selection problems. Results obtained from the procedure are validated with the results of previous researcher's results. It is concluded that the rank performance obtained from m-polar fuzzy ELECTRE-I integrated Shannon's entropy weight calculations method is similar to the similarity-based method and TOPSIS method for cutting fluid selection. In contrast, FMS selection ranks are identical to the COPRAS method, rank preference method. For the cutting fluid selection problem, alternative G is the best alternative. In the FMS selection problem, alternative C is the best alternative. Directed graphs were plotted to show the outranking relationships between other options. This new approach of implementing the m-polar fuzzy ELECTRE-I integrated Shannon's entropy weight calculations is helpful in the industrial selection problems. We can use this method for different industrial selection problems, like selecting rapid prototypes and selecting nontraditional machining processes. Researchers used the m-polar fuzzy ELECTRE-I algorithm to process multivalued input and evaluate the rank of alternatives. In the current study, we have taken single-valued information for the m-polar fuzzy ELECTRE-I method and validated results of rank performances with well-established MCDM methods.

References

1. Chen, J., Li, S., Ma, S., Wang, X., M-polar fuzzy sets: An extension of bipolar fuzzy sets. *Sci. World J.*, 2014, Article ID 416530, 8 pages, 2014.
2. Akram, M., Waseem, N., Liu, P., Novel approach in decision making with m-polar fuzzy ELECTRE-I. *Int. J. Fuzzy Syst.*, 21, 4, 1117–1129, 2019.
3. Prasad, K. and Chakraborty, S., Application of the modified similarity-based method for cutting fluid selection. *Decis. Sci. Lett.*, 7, 3, 273–286, 2018.
4. Rao, R.V. and Gandhi, O.P., Digraph and matrix method for the selection, identification and comparison of metal-cutting fluids. *Proc. Inst. Mech. Eng. Part J. J. Eng. Tribol.*, 215, 1, 25–33, 2001.
5. Sun, J., Ge, P., Liu, Z., Two-grade fuzzy synthetic decision-making system with use of an analytic hierarchy process for performance evaluation of grinding fluids. *Tribol. Int.*, 34, 10, 683–688, 2001.
6. Tan, X., Liu, D., Li, C., A decision-making framework model of cutting tool selection for green manufacturing and its application. *J. Adv. Manuf. Syst.*, 7, 2, 257–260, 2008.

7. Rao, R.V. and Patel, B.K., Decision making in the manufacturing environment using an improved PROMETHEE method. *Int. J. Prod. Res.*, 48, 16, 4665–4682, 2010.
8. Meciarova, J. and Stanovsky, M., Cutting fluids evaluation based on occupational health and environmental hazards. *Eng. Rural Dev.*, 10, 418–422, 2011.
9. Jagadish and A. Ray, Cutting fluid selection for sustainable design for manufacturing: An integrated theory. *Proc. Mater. Sci.*, 6, Icmpc, 450–459, 2014.
10. Deshamukhya, T. and Ray, A., Selection of cutting fluid for green manufacturing using analytical hierarchy process (Ahp): A case study. *Int. J. Mech. Eng. Robot. Res.*, 3, 1, 1–12, 2014.
11. Green cutting fluid selection using Moosra method. *Int. J. Res. Eng. Technol.*, 03, 15, 559–563, 2014.
12. Ravi Kumar, E.S.R. and Prasad, J.S.R., A novel approach of obtaining theoritical values in selection of cutting fluid attributes. *IOSR J. Math*, 10, 5, 01–04, 2014.
13. Tiwari, V.V., Madm for selection of vegetable based cutting fluids by saw method and wpm. *Int. J. Res. Technol. Manage.*, 1, 1, 16–28, 2015.
14. Prasad, K. and Chakraborty, S., A quality function deployment-based model for cutting fluid selection. *Adv. Tribol.*, 2016, Article ID 3978102, 10 pages, 2016.
15. Chatterjee, P. and Chakraborty, S., Flexible manufacturing system selection using preference ranking methods: A comparative study. *Int. J. Ind. Eng. Comput.*, 5, 2, 315–338, 2014.
16. Wabalickis, R.N., Justification of FMS with the analytic hierarchy process. *J. Manuf. Syst.*, 7, 3, 175–182, 1988.
17. Chan, F.T.S., Jiang, B., Tang, N.K.H., Development of intelligent decision support tools to aid the design of flexible manufacturing systems. *Int. J. Prod. Econ.*, 65, 1, 73–84, 2000.
18. Bayazit, O., Use of AHP in decision-making for flexible manufacturing systems. *J. Manuf. Technol. Manage.*, 16, 7, 808–819, 2005.
19. Rao, R.V., A decision-making framework model for evaluating flexible manufacturing systems using digraph and matrix methods. *Int. J. Adv. Manuf. Technol.*, 30, 11–12, 1101–1110, 2006.
20. Rao, R.V. and Parnichkun, M., Flexible manufacturing system selection using a combinatorial mathematics-based decision-making method. *Int. J. Prod. Res.*, 47, 24, 6981–6998, 2009.
21. Karande, P. and Chakraborty, S., Evaluation and selection of flexible manufacturing systems using MACBETH method. *Int. J. Serv. Oper. Manage.*, 16, 1, 123–144, 2013.
22. Shang, J. and Sueyoshi, T., A unified framework for the selection of a flexible manufacturing system. *Eur. J. Oper. Res.*, 85, 2, 297–315, 1995.
23. Sarkis, J., Evaluating flexible manufacturing systems alternatives using data envelopment analysis. *Eng. Econ.*, 43, 1, 25–47, 1997.

24. Talluri, S., Whiteside, M.M., Seipel, S.J., Nonparametric stochastic procedure for FMS evaluation. *Eur. J. Oper. Res.*, 124, 3, 529–538, 2000.
25. Karsak, E.E., Using data envelopment analysis for evaluating flexible manufacturing systems in the presence of imprecise data. *Int. J. Adv. Manuf. Technol.*, 35, 9–10, 867–874, 2008.
26. Liu, S.T., A fuzzy DEA/AR approach to the selection of flexible manufacturing systems. *Comput. Ind. Eng.*, 54, 1, 66–76, 2008.
27. Karsak, E.E., Distance-based fuzzy MCDM approach for evaluating flexible manufacturing system alternatives. *Int. J. Prod. Res.*, 40, 13, 3167–3181, 2002.
28. Rao, R.V., Evaluating flexible manufacturing systems using a combined multiple attribute decision making method. *Int. J. Prod. Res.*, 46, 7, 1975–1989, 2008.
29. Rao, R.V. and Singh, D., Evaluating flexible manufacturing systems using Euclidean distance-based integrated approach. *Int. J. Decis. Sci. Risk Manage.*, 3, 1/2, 32, 2011.
30. Jagtap, M. and Karande, P., Rank assessment of robots using m-polar fuzzy ELECTRE-I algorithm. *Proceedings of the International Conference on Industrial Engineering and Operations Management Bangalore*, India, August 16–18, 2021.

8

Fuzzy Decision Making: Concept and Models

Bithika Bishesh

Department of Management, Sharda University, Greater Noida, India

Abstract

Decision making is a critical aspect of human lives and is the most basic human activity wherein the best alternative is chosen from a set of given alternatives. The process of choosing from a set of alternatives is influenced by human subjectivity and, thus, depends on the value as well as the preferences of a decision maker. It is noteworthy that remaining neutral or deciding not to choose from any option from the given set of alternatives is also categorized as decision making. The field of decision making is useful in many fields, such as management, life sciences, engineering, and technology.

This chapter discusses the concept of fuzzy sets in decision making. Fuzzy sets were introduced in their modern form by Zadeh, wherein fuzzy sets were analyzed from the use of classical Boolean sets to a multivalued logic [1, 2]. The main idea behind fuzzy decision making is that the outcome is unknown, and the decision is made under uncertainty.

This chapter will give an overview of the use of fuzzy set theory in major types of decision-making problems. An introduction to fuzzy sets with their properties, extension, and membership function will also be discussed. Further, the multicriteria decision-making model will be discussed with a major emphasis on multiattribute decision making. Finally, some of the applications of fuzzy logic in sectors, such as agriculture will be discussed.

Keywords: Fuzzy logic, decision making, fuzzy sets, MADM

Email: bbishesh14@gmail.com

Sachi Nandan Mohanty, Prasenjit Chatterjee and Bui Thanh Hung (eds.) Fuzzy Computing in Data Science: Applications and Challenges, (147–172) © 2023 Scrivener Publishing LLC

8.1 Introduction

Decision making is a critical aspect of human lives and is the most basic human activity wherein the best alternative is chosen from a set of given alternatives. The process of choosing from a set of alternatives is influenced by human subjectivity and, thus, depends on the value, as well as the preferences of a decision maker. It is noteworthy that remaining neutral or deciding not to choose from any option from the given set of alternatives is also categorized as decision making. The field of decision making is useful in many fields, such as management, life sciences, engineering, and technology.

Mathematicians developed an interdisciplinary field called operations research that is focused on utilizing quantitative tools to solve problems and make choices accordingly. Mathematical tools and the application of scientific and technical approaches in the analysis of physical processes are central to operations research. Operations research provides the mathematical tools required for optimization. Formulating the problem statement is arguably the most challenging and critical aspect of any study in the process of operations research. The study's effectiveness hinges on the characterization of the problem, the identification of all variables that have a major impact on this problem, as well as selecting the results to be achieved. Additionally, for an operations research study to be effective, it must provide new knowledge about the given problem in a meaningful style to assist management in making more informed decisions.

Hence, fuzzy logic is a reasoning style that is very similar to human reasoning. This strategy is comparable to the way individuals make decisions. It also includes all choices between "yes" and "no." A set of the membership function is used for every input and every output to create a fuzzy logic system. The membership functions are then subjected to a set of rules in order to produce a "crisp" output value. Fuzzy logic is sometimes mistaken for probability theory. However, it is important to highlight that in fuzzy logic, we strive to convey the key concept of ambiguity, while probability is related to events rather than facts, and whether those events will occur or not.

The concept of fuzzy logic was once regarded as the theoretical philosophy of mathematicians with no possibility of application in the real world. This logic was conceptualized by Zadeh in 1965 [2]. Basically, it is an optimization technique. Most of the time, we try to gain maximum and lose minimum under any circumstances. However, sometimes, these circumstances act as constraints and restrict us in our target to achieve the maximum/minimum target. Through a process of optimization, we pay respect to and accept these constraints and then try to achieve what is possible between these two extremes. The resultant is called optimized value [3]. Say, for example, how a parent distributes his income to his children who may have varying levels of demands may not be entirely possible. Then the parent distributes his income in a way that satisfies "most" of the demands of his/her children. This is called optimization. We encounter two types of optimization:

When the demands (*variables* in mathematical terms) can be quantified to have "certain" values. Mathematical modelling of such problems is handled by Boolean logic [1].

When these demands are unable to be quantified and are "uncertain" in nature, then the mathematical modelling of such problems is handled by fuzzy logic. This model can be applied when situations are not clearly defined and thus have uncertainty. For example, categorizing people into high income, middle income, and low income is not completely clear, so an overlap of these categories may exist.

8.2 Classical Set

In a classical set, which is also known as the crisp set, a set is described to be a collection of well-defined distinct objects with a membership grade of either 0 or 1, depending on whether an element belongs to the set or not [4].

Representation of classic set A:

$$A = \{a_1, a_2, a_3 \ldots . a_n\}$$

where, A is the collection of all the elements $a_1, a_2, \ldots . a_n$.

$$A = \{x|P(x)\}, \text{ P: property}$$

The classical set uses Boolean logic as a characteristic function, which is dichotomous, which means yes or no, true or false, 0 or 1 type. The classical set is applicable only when the event is deterministic, the information is clear and ambiguous.

For instance, consider that a company is conducting a survey to collect product feedback, i.e., to find out whether its consumers are satisfied with the product or not. In the classical set or the Boolean logic aspect,

Figure 8.1 Comparison of Boolean logic with fuzzy logic.

the consumer's answer can either be 1 for yes or 0 for no. However, fuzzy logic mimics the human aspect of thinking so the answers will range from the customer is very satisfied, somewhat satisfied or not at all satisfied [5]. Hence, there are different values ranging from 0 to 1.

This distinction between fuzzy set and classical set is shown in Figure 8.1.

8.3 Fuzzy Set

If X is considered to be a universal nonempty set where x is an element of X, then a set A is called a fuzzy set on the set X if we can define it in the following manner [6]:

$$A = \{(x, \mu_A(x)):x\ X\}$$

Alternatively:

$$A= \left\{\left\langle \frac{\mu A(x)}{x} \right\rangle\right\} : x \in X$$

here,

A — universal set,

x — element on X,

$\mu_A : (x)$ — degree of membership or membership function.

A fuzzy is always represented by an ordered set of pairs, i.e., element of the set comes first in the pair followed by the degree of membership [7].

Membership function of fuzzy set A is defined as [8]:

$$\mu_A : X \rightarrow [0,1].$$

that is:

$$\mu_A(x) = \begin{cases} 1 \quad if\ x \in X \\ 0 \quad if\ x \notin X \\ (0,1) \qquad if\ x\ is \\ partially\ in\ X \end{cases}$$

8.4 Properties of Fuzzy Set

a) Empty Fuzzy set — any set A is called an empty fuzzy set if all the possible elements of that set have the value of membership function equal to zero. It can be represented as:

$$\mu_A(x) = 0$$

$$\text{Example A} = \left\{\frac{0}{a}, \frac{0}{b}, \frac{0}{c}\right\}$$

b) Universal Fuzzy set — any set A is considered a universal fuzzy set if the value of degree of membership is 1 for all the possible elements in that set. It can be represented as:

$$\mu_A(x) = 1$$

$$\text{Example A} = \left\{\frac{1}{a}, \frac{1}{b}, \frac{1}{c}\right\}$$

c) Height of Fuzzy set — largest value of the degree of membership of all the elements for any set A is known as the height of fuzzy set.

$$h(A) = \max_{x \in X} \{\mu_A(x)\}$$

Example: If there is a set A with the given degrees of membership

$$A = \left\{ \frac{0.3}{a}, \frac{0.6}{b}, \frac{0.8}{c} \right\}$$

then, height of fuzzy set, h(a) = 0.8.

d) Equality of 2 Fuzzy sets — two given fuzzy sets can either be equal or equivalent.

Equal fuzzy sets — consider there are two fuzzy sets that are represented as A and B. These are said to be equal fuzzy sets if these two sets have identical cardinality with same elements and all the elements have the same degree of membership [8]. Thus, A = B if,

$$A = \left\{ \frac{0.4}{a}, \frac{0.7}{b}, \frac{0.8}{c} \right\}$$

$$B = \left\{ \frac{0.4}{a}, \frac{0.7}{b}, \frac{0.8}{c} \right\}$$

Here, both these sets have the same elements (a, b, c) with the same membership function 0.4,0.7,0.8 respectively and $|A|=3$; $|B|=3$.

Equivalent fuzzy sets — consider there are two fuzzy sets A and B. These are said to be equivalent if both these sets have the same elements and cardinality but different membership functions. Thus, A B if,

$$A = \left\{ \frac{0.4}{a}, \frac{0.7}{b}, \frac{0.8}{c} \right\}$$

$$B=\left\{\frac{0.4}{a},\frac{0.6}{b},\frac{0.9}{c}\right\}$$

e) Containment of Fuzzy set — consider two sets A and B,

$$A=\left\{\frac{0.1}{a},\frac{0.3}{b},\frac{0.5}{c}\right\}$$

$$B=\left\{\frac{0.4}{a},\frac{0.6}{b},\frac{0.9}{c}\right\}$$

Both these sets have the same elements (a, b, c) and the same cardinality however, set A has lower values of membership function as compared to set B. Thus, set A is contained in B. This is called the containment of fuzzy sets.

8.5 Types of Decision Making

Fuzzy decision-making problems can be categorized in multiple ways. As per one approach, decision problems can be separated into two different categories: those that involve one single decision maker, as well as those that involve multiple decision makers [9, 10]. Individual decision making and multiperson decision making are the terms used to describe these problem types. The second approach differentiates decision-making problems involving multiple objective criteria. Moreover, decision making can also be accomplished in one or multiple phases/stages.

8.5.1 Individual Decision Making

Individual decision making is also called as single decision making since the decision is taken by a single person [11, 12]. In this type, the decision making problem is characterized as:

i) Set A of alternatives, $A = \{a_1,a_2,...\}$
ii) Set of criteria $C= \{C_1,C_2,...\}$
iii) Value of every alternative with respect to the given criteria $C_1(a_1),C_1(a_2)...$

The objective is to find a common solution out of these given alternatives. Suppose this fuzzy decision is represented by D then,

$$D = C_1 \cap C_2 \cap$$

In membership form

$$D\,(a_i) = \min\,\{C_1(a_i), C_2(a_i)....\} \quad (8.1)$$

Then, the most preferred decision is:

$$D\,(a^*) = \max\, D(a_i)$$

Alternatively,
Goals and constraints are represented as:

$$G_i(a) = G_{i'}(g_i(a)), \quad (8.2)$$

$$C_j(a) = G_{j'}(c_j(a)), \quad (8.3)$$

Using equations (8.2) and (8.3) we arrive at the final equation for decision making:

$$D\,(a) = \min\,[\inf G_i(a), \inf C_j(a)] \quad (8.4)$$

Example 1: Let us understand individual decision making with the help of an example. Suppose an individual receives three job offers a_1, a_2, and a_3. Then, the set of alternatives is:

$$A = \{a_1, a_2, a_3\}$$

While deciding to choose the best possible alternative out of the given set of alternatives, the individual finalises three criteria/parameters: salary, interesting job and future prospects, i.e., the number of years for the next promotion after entering the job. His goal is to get a high salary and there are two constraints—C_1 and C_2.

$$G_i = \text{high salary}$$

C_1= job is interesting

C_2= future prospects

For every criterion and each alternative, variables are shown in Table 8.1. We have to find out a suitable job that pays a high salary, is interesting, and with good future prospects.

Now, for each of these parameters, we have to define a membership value, which is alternatively known as the possibility value. It is the extent to which the element belongs to the fuzzy set.

$$\text{Salary} = \{(a_1, .65), (a_2, .70), (a_3, 1)\}$$

$$\text{Stability} = \{(a_1, 1), (a_2, .8), (a_3, .2)\}$$

$$\text{Future prospects} = \{(a_1, .3), (a_2, .8), (a_3, .8)\}$$

(for future prospects' membership grade calculation, it is assumed that the individual is willing to wait ten years for the next promotion, after entering the job. Thus, the formula used is

$$1 - \left(\frac{\text{number of years}}{\text{maximum number of years i.e.10}} \right)$$

Using equation (8.2) for composing functions $g_i(a)$ and G_i, we get the following value for salary (goal):

$$G_i = \frac{.65}{a1} + \frac{.7}{a2} + \frac{1}{a3}$$

This defines the goals in the context of set A's available jobs.

Table 8.1 Variables for individual decision making of job selection

Job needs	a_1	a_2	a_3
Salary G_i	65,000	70,000	100,000
Stability C_1	Very stable	Medium stability	Less stable
Future prospects C_2	7 years	2 years	2 years

Similarly, the fuzzy sets for the constraints stability (C_1) and Future prospects (C_2) are:

$$C_1 = \frac{1}{a1} + \frac{.8}{a2} + \frac{.2}{a3}$$

$$C_2 = \frac{.3}{a1} + \frac{.8}{a2} + \frac{.8}{a3}$$

Applying the formula for individual decision making, D (a) = min [inf-G_i(a), infC_j(a)],

For action a_1,

$$D\,(a_1) = \min\,[\inf G_i(a_1), \inf C_1(a_1)C_2(a_1)]$$

$$D\,(a_1) = \min\,[.65\ \inf(1,.3)]$$

$$D\,(a_1) = .3$$

For action a_2,

$$D\,(a_2) = \min\,[\inf G_i(a_2), \inf C_1(a_2)C_2(a_2)]$$

$$D\,(a_2) = \min\,[.7\ \inf(.8,.8)]$$

$$D\,(a_2) = .7$$

For action a_3,

$$D\,(a_3) = \min\,[\inf G_i(a_3), \inf C_1(a_3)C_2(a_3)]$$

$$D\,(a_3) = \min\,[1\ \inf(.2,.8)]$$

$$D\,(a_3) = .2$$

Final decision making value:

$$D(a) = \frac{.3}{a1} + \frac{.7}{a2} + \frac{.2}{a3}$$

Under the specified goal G with C1, and C2 as the given constraints, the most desired job out of the three jobs available is a2 since it has the highest membership grade of 0.7.

8.5.2 Multiperson Decision Making

Whenever a group decision is to be formed on the basis of the individual decision of all members of the group, it may happen that individual decision makers may differ in selecting the alternatives [13]. In order to deal with the different varying decisions of group members, some effective methodology has to be devised for aggregating all the decisions of the group members. In the simplest form,

Let, $S(x_i, x_j)$ be defined as a relative acceptance of alternative x_i over x_j to be calculated as,

= (Number of individuals favoring x_i over x_j) / (the total decision makers, n)

Mathematically, it can be rewritten as,

$$S(x_i, x_j) = N(x_i, x_j) / n \tag{8.5}$$

Example 2: Let us revisit the individual decision making problem in Example 1. Now a group of six individuals want to have a job change together in another company. Every person in this group has total preference ordering A_i (i N_6) on a set of alternatives $X = \{a_1, a_2, a_3\}$ as follows:

$A_1 = \{a_1, a_2, a_3\}$ $A_2 = \{a_1, a_3, a_2\}$ $A_3 = \{a_2, a_1, a_3\}$

$A_4 = \{a_2, a_3, a_2\}$ $A_5 = \{a_3, a_1, a_2\}$ $A_6 = \{a_3, a_2, a_1\}$

We have the membership function represented as,

$$S(x_i, x_j) = N(x_i, x_j) / n$$

Taking n equal to 6 and using the equation we can derive the fuzzy social preference relation as follows:

$$S = \begin{array}{c} \\ a_1 \\ a_2 \\ a_3 \end{array} \begin{array}{c} \begin{array}{ccc} a_1 & a_2 & a_3 \end{array} \\ \begin{bmatrix} 0 & 0.5 & 0.625 \\ 0.5 & 0 & 0.375 \\ 0.25 & 0,25 & 0 \end{bmatrix} \end{array}$$

After defining the fuzzy relationship S, it can be converted to its resolution form in order to find out the final nonfuzzy group preference.

$$S = \bigcup_{\alpha \in [0,1]} \alpha^{\alpha} S$$

This is a union of the crisp relations ${}^{\alpha}S$ which consists of α-cuts of S, where each is scaled by α. The largest value α for any group shows the maximized agreement level for the group and represents group decision [14].

8.5.3 Multistage Decision Making

Multistage decision making shows N-variable problems as N single-variable problems [15]. The optimum solutions of all the N single variable problems are used for finding out the optimum value of the actual problem by solving the problems in a sequential manner. Single variable problems are linked in a series in order to ensure that the output of one stage becomes the input of the next. This specific type of problem is known as a serial multistage decision process [16].

Consider the problem of allocating irrigation water to N number of farmers. In this example, maximizing the total net benefit received by all the farmers becomes our objective function. To address this problem, every farmer has to be taken individually and then optimize their individual net benefits, which will be subject to constraints. After this, the sum total of the benefit received by all the farmers will be calculated to find out the total optimal benefit.

Suppose an automation with $X = (x_1, x_2)$, $Z = (z_1, z_2, z_3)$ as well as state-transition function be represented by the given matrix,

$$\begin{array}{c} \\ z_1 \\ z_2 \\ z_3 \end{array} \begin{array}{c} \begin{array}{cc} x_1 & x_2 \end{array} \\ \begin{bmatrix} z_1 & z_2 \\ z_3 & z_1 \\ z_1 & z_3 \end{bmatrix} \end{array}$$

Let us consider $N = 2$ and, fuzzy goal at $t = 2$ as

$$C^2 = 0.3/\check{z}_1 + 1/z_2 + 0.8/z_3$$

It is further assumed that, the given fuzzy constraints for inputs at time $t=0$ and time $t=1$ are,

$$A^0 = 0.7/x_1 + 1/x_2,$$

$$A^1 = 1/x_1 + 0.6/x_2.$$

For solving this decision problem, a sequence $\hat{x}^0$, $\hat{x}^1$ of input states for which maximum is obtained of the expression,

$$\max_{x^0,\ x^1} \min\left[A^0\left(x^0\right), A^1\left(x^1\right), C^2\left\{f\left(z^1, x^1\right)\right\}\right]$$

By using the first backward iteration for time

$t = 1$, we get $C^1(z_1)=0.6$, $C^1(z_2)=0.8$, $C^1(z_3)=0.6$

from where we get $C^1 = 0.6/z_1 + 0.8/z_2 + 0.6/z_3$
and, after maximization of equation, $\min\left[A^1\left(x^1\right), C^2\left\{f\left(z^1, x^1\right)\right\}\right]$

The optimal decision $\hat{x}^1$ for every state z_1 Z at $t = 1$ is written as,

z^1	z_1	z_2	z_3
$\hat{x}^1$	x_2	x_1	x_2

for time $t = 0$ in the second backward iteration, we get $C^0(z_1)=0.8$, $C^0(z_2)=0.6$, $C^0(z_3)=0.6$ from where we get $C^0 = 0.8/z_1 + 0.6/z_2 + 0.6/z_3$
and, after maximization of equation, $\min\left[A^0\left(x^0\right), C^1\left\{f\left(z^0, x^0\right)\right\}\right]$

The optimal decision $\hat{x}^0$ for every state z_0 Z at $t = 0$ is,

z^0	z_1	z_2	z_3
$\hat{x}^0$	x_2	x_1	x_2

It means that when our initial state is given as z_1, then the maximizing decision is of applying the action x_2 first and, then action x_1. Here in the given case, our goal is found to be satisfied to a degree $C^0(z_1) = 0.8$.

8.5.4 Multicriteria Decision Making

This is a method for selecting the best option from an array of choices based on a set of criteria [17].

Thus, in this type of decision making problem, the following two steps are involved:

a) In the first step, the selection of a number of alternatives in limited quantities is done by a method known as multiattribute decision making (MADM) which involves the evaluation of alternatives on the basis of the various attributes of the objective. These attributes could be numerical data, demographical data, historical data, qualitative data and likewise.
b) In the second step, the best alternative is determined using multiobjective decision making (MODM).

Mathematically, a universe of n alternatives can be defined as,

$$A = (a_1, a_2, a_3, ...a_i, ...a_n)$$

and, the set of m objectives as,

$$O = (o_1, o_2, o_3, ...o_i, ...o_n)$$

$$R = \begin{array}{c} \\ o_1 \\ O_2 \\ \cdots \\ \\ O_n \end{array} \begin{array}{c} \begin{array}{cccc} a_1 & a_2 & \cdots & a_n \end{array} \\ \begin{bmatrix} r_{11} & r_{12} & \cdots & r_{1n} \\ r_{21} & r_{22} & \cdots & r_{2n} \\ \cdots & \cdots & \cdots & \cdots \\ r_{m1} & r_{m2} & \cdots & r_{mn} \end{bmatrix} \end{array}$$

Suppose that all entries r_{ij} given in the above matrix are considered to be real numbers [0,1]. Further, assume that every entry shows the extent of the criterion c_i being satisfied by the alternative x_j (i N_m, j N_n). R can then be thought of as the matrix depiction of the fuzzy relationship on O x A.

It is to be noted that, the situation may arise where, in place of matrix R with the entries [0,1] another alternative matrix is given such that R' = $[r'_{i,j}]$ is given with random real number entries. In this scenario, the formula that we can use to transform R' to R is:

$$r_{ij}^{'} = \frac{r_{ij}^{'} - \min_{j \in N_n} r_{ij}^{'}}{\max_{j \in N_n} r_{ij}^{'} - \min_{j \in N_n} r_{ij}^{'}} \tag{8.7}$$

Multiobjective decision problems are converted to single-objective decision problems by finding out the global criterion, $r_j = h(r_{1j}, r_{2j}, \ldots, r_{mj})$ [18].

Thus, multicriteria decision making is of two main types:

1. Multiobjective decision making—It is a methodical approach to making decisions that enables a decision maker to describe a decision, establish their individual value system, contextualize the situation, and assess the conclusion in a systematic manner. The analyst can break down a complex decision into smaller and simpler components by establishing the decision alternatives (what possibilities are available) as well as the decision criteria (what is relevant to the decision maker) [19]. These given alternatives are ranked against the decision criteria after the project context and structure have been established. The criteria are then weighted based on the value structure of the decision maker, yielding a final score allocation to the different decision alternatives [20].
2. Multiattribute decision making—It entails developing preference-based decisions among the available options with various, often contrasting features. Multiattribute decision-making issues are numerous and can be found in almost any subject [21]. They are used to determine which of multiple comparable alternatives is the most acceptable, or to rank alternatives as per their relevance to the evaluated objective. These methods are used to choose the most satisfactory solution if there is no such alternative with the best values for all criteria. When using traditional optimization methods to address a problem, the function of the goal is fixed, defining the collection of objects to be optimized or the permissible area to be defined [22]. Depending on the goal of the task, the minimum or the maximum values of a function are chosen.

8.6 Methods of Multiattribute Decision Making (MADM)

Multiattribute decision-making methods have been effectively used in different decision-making situations. They have the ability to consider both technical and nontechnical factors while making decisions. Several approaches have been created based on application areas of decision making. Below are brief descriptions of some of the commonly employed methods for evaluating various solutions for real-world applications.

8.6.1 Weighted Sum Method (WSM)

The Weighted Sum Method (WSM) is a very basic method, with applications in addressing problems that are single dimensional. It considers every alternative's total value as the same as the total value of the products. Because of its simplicity, it is difficult to be applied to those problems where units of criteria (attributes) are not the same and differ largely.

The following equation yields the best answer to the problem when there are n criteria and m alternatives:

$$A^*_{WSM} = \max_{i} \sum_{j=1}^{n} a_{ij} w_j \tag{8.8}$$

here $i = 1,.......m,$

$$A^*_{WSM} \text{ — weighted sum score,}$$

$$a_{ij} \text{ — score of } i\text{-th alternative for } j\text{-th criterion}$$

$$w_j \text{ — weight of } j\text{-th criterion.}$$

8.6.2 Weighted Product Method (WPM)

The Weighted Product Method involves a minor variation in the WSM methodology discussed above. The primary difference between these two is that WPM uses a product instead of a sum [23]. Each option is evaluated with the others using multiplication of ratios associated with each criterion.

$$A^{*}_{WSM} = \underset{i}{max} \prod_{j=1}^{n} (r_{ij}^{*})^{w_j} \tag{8.9}$$

WPM has been found to be useful in both single-dimensional and multidimensional instances. In the equation below, this approach compares the alternatives a_k and a_l. While comparing in pairs, the best solution is one having at least a similar value to the remaining alternatives, and in a maximization situation, the best option is a_k when $R\left(\frac{A_k}{A_l}\right) > 1$

$$R\left(\frac{A_k}{A_l}\right) = \prod_{j=1}^{n} \left(\frac{a_{kj}}{a_{lj}}\right)^{2} \tag{8.10}$$

8.6.3 Weighted Aggregates Sum Product Assessment (WASPAS)

Weighted Aggregates Sum Product Assessment combines the approaches of the Weighted Sum Model as well as the Weighted Product Model. The WASPAS is a compensating method in which the attributes are not dependent on one another and qualitative attributes can be transformed into quantitative attributes [24].

The WASPAS approach is utilized to rank alternatives in risk assessment and selection of the maintenance strategy, while Standard Deviation analyzes the weights of decision criteria.

a) Standard Deviation (SD) Approach

The calculation of decision criteria weights is a crucial part of the overall process of the decision-making process since criteria weights have a significant impact on the ranking of the alternatives. This is accomplished using one of the decision-making techniques called as the Standard Deviation technique. This method determines relative relevance so that the attribute with greater deviation is more important than those having smaller deviation [25].

The following steps are included in this type of decision-making problem:

Table 8.2 Creating a decision matrix.

Alternative (A_i)	Decision criterion (C_j)			
A_1	$x_{1,1}$	$x_{1,2}$	—	$x_{1,n}$
A_2	$x_{2,1}$	$x_{2,2}$	—	$x_{2,n}$
—	—	—	—	—
A_m	$x_{m,1}$	$x_{m,2}$	—	$x_{m,n}$

a) Creating a decision matrix
The first step in determining the weight of the decision criteria is to create a decision matrix as given in Table 8.2.
Here, x_{ij} is the rating given to the alternative i in terms of j-th decision criteria.

b) Normalization of decision matrix
Normalization is done using the formula:

$$r_{ij} = \frac{x_{ij}}{\sqrt{\sum_{j=1}^{m} x_{ij}}} \tag{8.11}$$

r_{ij} — normalized decision matrix.

c) Evaluating the performance index on the basis of WSM and WPM
Standard deviation is analyzed separately for every decision criterion:

$$sV_j = \sqrt{\frac{1}{m}\sum_{i=1}^{m}\left(r_{ij} - \bar{r}_j\right)^2} \tag{8.12}$$

$\bar{r}_j$ — mean value of j-th decision criterion
sV_j — standard deviation for j-th decision criterion.

d) Finding the Weight
The decision criteria weights are calculated as:

$$W_j = \frac{sV_j}{\sum_{i=1}^{n}\left(sV_j\right)} \tag{8.13}$$

b) WASPAS Approach

a) Normalization of decision matrix
Normalization is done using the formula:

$$p_{ij} = \frac{x_{ij}}{\max\limits_{i} x_{ij}}, \quad j = 1, 2, \ldots n, \quad i = 1, 2, \ldots m \tag{8.14}$$

b) Evaluating the performance index on the basis of WSM and WPM
Performance index of i-th alternative by WSM is calculated as:

$$sQ_i = \sum_{j=1}^{n} p_{ij}.w_j \tag{8.15}$$

Performance index of i-th alternative by WPM is calculated as:

$$pQ_i = \prod_{j=1}^{n} \left(p_{ij}\right)^{w_j} \tag{8.16}$$

For a typical decision matrix, Equations 8.15 and 8.16 are systematically combined to build one performance model, which is used for ranking of alternatives:

$$Z = 0.5sQ_i + pQ_i = 0.5\sum_{j=1}^{n} p_{ij}.w_j + 0.5\prod_{j=1}^{n}\left(p_{ij}\right)^{w_j} \tag{8.17}$$

$$Z = \lambda\, sQ_i + (1-\lambda)\, pQ_i = \lambda \sum_{j=1}^{n} p_{ij}.w_j + (1-\lambda)\prod_{j=1}^{n}\left(p_{ij}\right)^{w_j} \tag{8.18}$$

8.6.4 Technique for Order Preference by Similarity to Ideal Solutions (TOPSIS)

The idea behind this approach is that the alternative that is selected should have the shortest distance from fuzzy positive ideal solution (FPIS) and longest distance from fuzzy negative ideal solution (FNIS). The normalized decision matrix is created after establishing n criteria and m alternatives [26].

Normalized value r_{ij} is shown in equation (8.19).

$$r_{ij} = \frac{f_{ij}}{\sqrt{\sum_{j=1}^{m} f_{ij}}^{2}} \tag{8.19}$$

$$j = 1, \ldots, m$$

$$i = 1, \ldots, n$$

f_{ij} - Value of i-th for the alternative A_j

For the given decision matrix, we calculate the normalized weighted values v_{ij}:

$$v_{i\ j} = w_i r_{i\ j} \tag{8.20}$$

A^-, the fuzzy positive ideal solution is:

$$A^+ = \{v_1 +, \ldots, v_n +\} = \{(MAX_j v_{ij} \text{ where } i \in I'), (MIN_j v_{ij} \text{ where } i \in I'')\} \tag{8.21}$$

A^- the fuzzy negative ideal solution is:

$$A^- = \{v_1 +, \ldots, v_n +\} = \{(MIN_j v_{ij} \text{ where } i \in I'), (MAX_j v_{ij} \text{ where } i \in I'')\}$$

I' and I" represent the benefit and the cost criteria (i.e., the positive and the negative variables).

Equation (8.23) shows the calculation for separation of each of the alternatives from A^+ and equation (8.24) shows the calculation for separation of each of the alternatives from A^-.

$$D_j^+ = \sqrt{\sum_{i=1}^{n} \left(v_{ij} - v_i^+\right)^2} \tag{8.23}$$

$$D_j^- = \sqrt{\sum_{i=1}^{n} \left(v_{ij} - v_i^-\right)^2} \tag{8.24}$$

Comparative closeness to the given ideal solution for every alternative is:

$$C_j = \frac{D_j^-}{\left(D_j^+ + D_j^-\right)} \tag{8.25}$$

Once the Cj values are sorted, then the optimal solution will be the corresponding maximum value.

8.7 Applications of Fuzzy Logic

Beginning the second half of the last century, a lot of research has been done worldwide and researchers have found numerous scope of fuzzy logic applications. Its application ranges from managerial decisions to technical innovations, from scientific discoveries to robotics, from the field of humanities to the field of biological and medical sciences and in many more areas. In fact, you think of any field and literature and applications are available to highlight applications in that field. Fuzzy programming has proved to be beneficial in several areas such as optimization, robotics, automation, image processing, consumer electronics, and power and control systems engineering.

Fuzzy logic with its human decision making can solve various problems of agriculture and thus aid in its development. One of the basic needs of agriculture is the suitability of land and climate conditions necessary for a crop. Farmers traditionally rely on the knowledge inherited by them from generation to generation. Fuzzy logic systems have also been used in the context of climate prediction for agriculture [27]. Fuzzy logic application has also been devised for assessing soil quality and suitability for agricultural purposes. The quality of soil is predicted more accurately by fuzzy logic reasoning in soil mapping [28].

Fuzzy logic systems have also been devised to develop decision support systems for supply-chain management [29] as it is affected by several controllable variables, like perishability and season based cultivation in the agribusiness and agroindustrial domains are affected by several controllable variables, like perishability and season based cultivation.

To manage and mitigate pests and weeds is another challenge faced by the agricultural industry. In this area, fuzzy systems in tandem with computer techniques have come to the rescue of the farmers by more effectively and efficiently controlling pests and weeds. It can calculate and predict the risk of likely diseases in plants for which risk factors and symptoms are considered.

There are many other areas where fuzzy logic has been successfully used. They are as briefly stated below:

- Fuzzy logic application has enabled the design of brakes to be applied in perilous cases depending on acceleration and, wheel speed.
- It is also employed in automotive fuel injection and ignition controls depending on throttle position, revolutions per minute, cooling the water temperature and, other factors.
- Design of automatic transmission by autoselection of appropriate gear as per engine load and road conditions.

Interestingly, besides engineering applications, the fuzzy logic application has also found its entry into the field of humanities and life sciences. Some of the key areas are as described below:

- Business—to evaluate employees and formulate support systems that facilitate decision making [30].
- Defence—for identifying targets automatically through infrared thermal images, detecting an underwater target, and developing support aids for naval decisions.
- Finance—to control banknote transfer, manage the fund and develop systems for stock market predictions and securities trading.

- Medical—for the medical diagnostic support system, multivariable controlling of anesthesia and, radiology diagnostics.
- Pattern recognition—for speech recognition handwriting recognition, analyzing facial characteristics, etc.
- Psychology—for criminal profiling and analyzing human behavior in general.

8.8 Conclusion

Fuzzy decision making is a reasoning style that is very similar to human reasoning. This strategy is comparable to the way individuals make decisions. Beginning the second half of the last century, a lot of research has been done worldwide, and researchers have found numerous scopes of fuzzy logic applications. Fuzzy logic with its human decision making can solve various problems in different areas, like optimization, robotics, automation, image processing, consumer electronics, and power and control systems engineering. Thus, its application ranges from managerial decisions to technical innovations, from scientific discoveries to robotics, from the field of humanities to the field of biological and medical sciences, and in many more areas.

References

1. Blanco-Mesa, F., Merigó, J.M., Gil-Lafuente, A.M., Fuzzy decision making: A bibliometric-based review. *J. Intell. Fuzzy Syst.*, 32, 3, 2033–2050, 2017.
2. Zadeh, L.A., Fuzzy sets. *Inf. Control*, 8, 394–432, 1965.
3. Sunardi, S., similarity-sunardi-0521057401-C. 1, in: *14-Taxonomy of fuzzy multi attribute decision making.*
4. Sivanandam, S.N., Sumathi, S., Deepa, S.N., *Introduction to fuzzy logic using MATLAB*, vol. 1, Springer, Berlin, 2007.
5. Baharvand, S., Rahnamarad, J., Soori, S., Saadatkhah, N., Landslide susceptibility zoning in a catchment of Zagros Mountains using fuzzy logic and GIS. *Environ. Earth Sci.*, 79, 10, 1–10, 2020.
6. Rao, S.S., *Engineering optimization: theory and practice*, pp. 722–727, John Wiley & Sons, 2019.
7. Skalna, I., Rebiasz, B., Gawel, B., Basiura, B., Duda, J., Opila, J., Pelech-Pilichowski, T., Advances in fuzzy decision making. *Stud. Fuzziness Soft Comput.*, 333, 27–46, 2015.
8. Zimmermann, H.J., *Fuzzy set theory—and its applications*, Springer Science & Business Media, 2011.

9. Shakhovska, N. and Stepashko, S., Advances in intelligent systems and computing II: Selected papers from the *International Conference on Computer Science and Information Technologies, CSIT 2017*, September 5–8 Lviv, Ukraine, 2017.
10. Sayyaadi, H. and Esmaeilzadeh, H., Determination of optimal operating conditions for a polymer electrolyte membrane fuel cell stack: Optimal operating condition based on multiple criteria. *Int. J. Energy Res.*, 37, 14, 1872–1888, 2013.
11. Mei, S., Zhu, Y., Qiu, X., Zhou, X., Zu, Z., Boukhanovsky, A.V., Sloot, P.M., Individual decision making can drive epidemics: A fuzzy cognitive map study. *IEEE Trans. Fuzzy Syst.*, 22, 2, 264–273, 2013.
12. Merigó, J.M., Fuzzy decision making with immediate probabilities. *Comput. Ind. Eng.*, 58, 4, 651–657, 2010.
13. Cabrerizo, F.J., Chiclana, F., Al-Hmouz, R., Morfeq, A., Balamash, A.S., Herrera-Viedma, E., Fuzzy decision making and consensus: Challenges. *J. Intell. Fuzzy Syst.*, 29, 3, 1109–1118, 2015.
14. Kacprzyk, J. and Fedrizzi, M., *Multiperson decision making models using fuzzy sets and possibility theory*, Kluwer academic publ., pp. 231–241, 1990.
15. Sirbiladze, G., Khutsishvili, I., Ghvaberidze, B., Multistage decision-making fuzzy methodology for optimal investments based on experts' evaluations. *Eur. J. Oper. Res.*, 232, 1, 169–177, 2014.
16. Chen, S.Y. and Fu, G.T., Combining fuzzy iteration model with dynamic programming to solve multiobjective multistage decision making problems. *Fuzzy Sets Syst.*, 152, 3, 499–512, 2005.
17. Pedrycz, W., Ekel, P., Parreiras, R., *Fuzzy multicriteria decision-making: Models, methods and applications*, pp. 103–106, John Wiley & Sons, 2011.
18. Kahraman, C., Onar, S.C., Oztaysi, B., Fuzzy multicriteria decision-making: A literature review. *Int, J. Comput. Intell. Syst.*, 8, 4, 637–666, 2015.
19. Kentel, E. and Aral, M.M., Fuzzy multiobjective decision-making approach for groundwater resources management. *J. Hydrol. Eng.,* 12, 2, 206–217, 2007.
20. Brevault, L., Pelamatti, J., Hebbal, A., Balesdent, M., Talbi, E.G., Melab, N., MDO related issues: Multi-objective and mixed continuous/discrete optimization, in: *Aerospace system analysis and optimization in uncertainty*, pp. 321–358, Springer, Cham, 2020.
21. Li, D.F., A fuzzy closeness approach to fuzzy multi-attribute decision making. *Fuzzy Opt. Decis. Making*, 6, 3, 237–254, 2007.
22. Ren, Z. and Wei, C., A multi-attribute decision-making method with prioritization relationship and dual hesitant fuzzy decision information. *Int. J. Learn Mach. Cybern.*, 8, 3, 755–763, 2017.
23. Mateo, J.R.S.C., Weighted sum method and weighted product method, in: *Multi criteria analysis in the renewable energy industry*, pp. 19–22, Springer, London, 2012.

24. Chakraborty, S., Zavadskas, E.K., Antucheviciene, J., Applications of WASPAS method as a multi-criteria decision-making tool. *Econ. Comput. Econ. Cybern. Stud. Res.*, 49, 1, 5–22, 2015.
25. Mardani, A., Nilashi, M., Zakuan, N., Loganathan, N., Soheilirad, S., Saman, M.Z.M., Ibrahim, O., A systematic review and meta-analysis of SWARA and WASPAS methods: Theory and applications with recent fuzzy developments. *Appl. Soft Comput.*, 57, 265–292, 2017.
26. Pavić, Z. and Novoselac, V., Notes on TOPSIS method. *Int. J. Res. Eng. Sci.*, 1, 2, 5–12, 2013.
27. Marimin, M. and Mushthofa, M., Fuzzy logic systems and applications in agro-industrial engineering and technology, in: *2nd International Conference on Adaptive and Intelligent Agroindustry*, 2013, September.
28. Qi, F., Zhu, A.X., Harrower, M., Burt, J.E., Fuzzy soil mapping based on prototype category theory. *Geoderma*, 136, 3-4, 774–787, 2006.
29. Dweiri, F.T. and Kablan, M.M., Using fuzzy decision making for the evaluation of the project management internal efficiency. *Decis. Support Syst.*, 42, 2, 712–726, 2006.
30. Afshari, R.A., Nikolić, M., Ćoćkalo, D., Applications of fuzzy decision making for personnel selection problem: A review. *J. Eng. Manage. Competitiveness (JEMC)*, 4, 2, 68–77, 2014.

9

Use of Fuzzy Logic for Psychological Support to Migrant Workers of Southern Odisha (India)

Sanjaya Kumar Sahoo and Sukanta Chandra Swain*

KIIT Deemed to be University, Bhubaneswar, India

Abstract

People migrate with lot of ambitions to turn their economic status around. In Odisha, particularly in the southern part of the Odisha state, there are many families where the migrant worker is the only earning member in the family and also migration is the only option to choose. Even if the government and NGOs have taken so many steps for the betterment of the migrant workers, improper way of working continues and also migrant workers are being taken for granted. The pandemic caused by COVID-19 has shattered the dreams of migrant workers as they have been the hardest hit of the virus both from lives and livelihoods fronts. This paper is to present the psychological status of migrant workers during COVID-19 and for the purpose, the migrant workers of two southern districts (Khurdha and Gajapati) of Odisha have been considered. Moreover, how fuzzy logic can be of help in improving the psychological state of such migrants has also been highlighted in this paper.

Keywords: Fuzzy logic, COVID-19, effect, psychological status, migrant workers, Odisha

**Corresponding author*: sukanta_swain@yahoo.com

Sachi Nandan Mohanty, Prasenjit Chatterjee and Bui Thanh Hung (eds.) Fuzzy Computing in Data Science: Applications and Challenges, (173–180) © 2023 Scrivener Publishing LLC

9.1 Introduction

People migrate to occupy a better socioeconomic status in the society by tolerating so many difficulties like getting treated as outsider, deprived of getting proper hygienic food, leaving in a unhealthy atmosphere, fear, anxiety, staying outside the home, etc. Despite all these, migrated people adjust themselves with the new environment to earn something to revive their family from the economic backwardness. In fact, dreams like construction of their houses in the village, marriage of their sister, financing education of their children, saving the parents to work in old age, maintaining a descent standard of living, etc. are the cause of migration from rural Odisha. This type of ambition makes every odd at the end of each working day a sleigh of solace. It is because they were mentally strong and determined to get their objectivity not only for themselves but also for their dependent family members. The pandemic COVID-19 affected almost all the people of the world in general and most severely the migrant workers in particular. Now their dream changed from highly ambitious to saving their life. Pandemic COVID-19 brings such a situation that all the employers give direction to their employees to halt every productive activity due to lock down. Further, they directed their workers to move away from the productive place to their native place even though the government instructed them not to do like this. Migrants, who are considered as the backbone of the economic development of the country, have now become a burden to every sector. In this critical juncture, most of the migrant labor move to their native place miles and miles kilometer even by walking, cycling, boat, etc. in the mid-way to their native place, some died and some conceived migrant lady give birth to baby. Realizing their pathetic situation, different organizations like the government of India, all the state governments, NGOS, philanthropist individuals like Sonu Sood and Sabyasachi Mishra helped them in rescuing. Even though they reached their native places, they were not allowed to enter their villages rather they were to stay in quarantine for 21 days. In India, as per NSS report, 28.3% of the labor force were migrant in 2007–2008 [9]. As per the census of India, there were 454 million, 315 million, and 220 million migrants as per census 2011, 2001, and 1991, respectively [10]. In this situation, the paper tries to explore the psychological and emotional effect of COVID-19 on migrant labor.

It is found that internal migrants are severely affected due to COVID-19. Despite so many steps taken, COVID-19 affects them psychologically. Mental health community needs to take steps for handling these things [1]. Lockdown for COVID-19 has influenced the health of the migrants.

With the deterioration of mental health due to losing job, insecurity may continue even during the post lockdown period. Poor management of the health service may lead to increase in suicidal activity. So necessary steps are necessary wherein people can get mental health service [2]. Migrants are living in depression and anxiety. Besides, negative emotions like loneliness, tension, low mood, fear, irritability, etc. have also been developed among them [3]. Mental health that characterizes migration gets compromised by a series of interacting social and environmental variables [4]. It is found that migrant workforce reveal more psycho, temper, anxiety disorders, and posttraumatic disorders [5]. The risk increases for first-generation migrants. The most important risk factors include failure to complete projects, cross-cultural stress, and loss of social status, discrimination, and separations from the family context [6]. The presence of compatriots, on the other hand, is a protective factor, even though it can sometimes hinder integration into the community, threatening long-term adaptation. Fuzzy cognitive maps (FCMs) is a soft computing know-how for modeling human cognition and developing quantitative dynamic models [7]. A model based on cognition, like FCM, presents a plan of action in line with the way humans comprehend it. Hence, the model is well comprehensible, even by an amateur mass, and each factor has a perceptible meaning. It can be easily made flexible to integrate new attributes. Thus, an FCM is an impulsive tool for modeling in which the resoluteness of the scheme content can be augmented by following an extended mapping [8].

Thus, FCM can be a tool for ascertaining the psychological issues experienced by the migrants and devise the supportive steps for coming out of that.

9.2 Objectives and Methodology

9.2.1 Objectives

The objectives of the chapter are as follows.

- To ascertain the effect of COVID-19 on the psychology and emotion of repatriated migrants of Khurdha and Gajapati districts of Odisha.
- To suggest the way out for strengthening the psychological strength of the migrant workers during this difficult situation by way of using fuzzy logic.

9.2.2 Methodology

The methodologies adopted for the purpose are follows:

- To explore the psychological and emotional effect of COVID-19 on the repatriated migrants of Khurdha and Gajapati districts of Odisha, primary data are collected through telephonic conversation from 40 repatriated migrants of Khurdha and Gajapati districts of Odisha. Snow ball sampling method is adopted to identify the targeted respondents. A semistructured questionnaire is administered for this purpose. To ascertain the way they manage their time, migrants are segregated into educated and uneducated.
- Experts' opinion are considered for suggesting the way out for strengthening the psychological strength of the migrant workers through technology during these difficult situations.

9.3 Effect of COVID-19 on the Psychology and Emotion of Repatriated Migrants

9.3.1 Psychological Variables Identified

To know the psychological and emotional effect of the migrant workers, different variables have been taken and their response mentioned in Table 9.1.

Table 9.1 shows that 14 variables have taken to know the psychological and emotional effect of COVID-19 on the migrant workers. The response shows that these variables have increasing impact among the maximum people. Very few people show that they do not see any change. Similarly, very few people responded that these variables have decreasing impact among them.

9.3.2 Fuzzy Logic for Solace to Migrants

For knowing the status of migrants on the scale of poverty, fuzzy logic can be used. For the usage of fuzzy logic, let us presume that there exists a poverty line wP and a richness line wR ($wP<wR$). The t-th case is considered poor (non-rich) if $wt<wP$. In a similar way, we say that the t–th case is rich (non-poor) when $wt \geq wR$. Finally, when $wP \leq wt < wR$, we can conclude that the t–th case belongs to an obscurity related to the non-poor and non-rich case. Hence, since the t–th case cannot be considered rich, it follows that such an observation is poor *to some extent*. Therefore, we no longer

Table 9.1 Variables considered for psychological and emotional effect of the migrant workers.

Variables	No change (in %)	Increase (in %)	Decrease (in %)	Total (in %)
Loneliness	5	85	10	100
Irritation	8	70	22	100
Frustration	3	90	7	100
Upset	5	95	0	100
Death fear (self)	5	80	15	100
Tension	10	85	5	100
Believe in God	30	35	35	100
Loss of job fear	10	60	30	100
Death fear (family member)	5	80	15	100
Watching TV	20	65	15	100
Use of Social Media	15	75	10	100
Fear of watching any outsider as the spreader of disease	35	55	10	100
Attempt to take medicine for precaution	25	60	15	100
Anxiety to know about the findings of Vaccine/Medicine	20	70	10	100

Source: Primary data

distinguish poor and rich (non-poor) observation units in a dichotomous way. Instead, pursuing the fuzzy logic formulation, we allot to every case a degree of poverty ranging from 0 to 1 according to the related income. This can be done by pioneering the so-called membership function (of a case t to the attribute "poor"). In particular, if $W\sim$ indicates the attribute "poor" (the symbol "~" denotes that a fuzzy logic approach is adopted), the membership function μ $W\sim$: R+ → [0,1] allows us to express to which extent the *t*–th case is poor. This is done by quantifying the degree according to which *wt* belongs to $W\sim$.

As we could identify the status of the migrants through the membership function, by way of direct personal interview and in-depth interviews, we could also study the psychological status of each of those migrants. Then by conducting focus group discussions among experts and psychologists, we could also trace out the tentative solution for their problems.

9.4 Findings

Following are the findings of the study.

- It demoralized the people to work outside the native place.
- Most of the people thought that if God exists, then why did the virus come out? And very few people thought that even though virus came people are saved because of God.
- It increased the use of social media.
- It increased the anxiety to know when the vaccine or medicine will come out.
- It increased the frustration among the people.
- Most of the people search to medicine store to know as if some medicines are available for precaution purpose.
- It makes the people more irritate.
- During the lockdown period, people feel loneliness.
- It creates a psychological fear that death is now knocking on the door.
- It creates anxiety about the safety of a family member.
- Most of the people fear of losing their job.
- Those who have TV spend their time by watching it.

9.5 Way Out for Strengthening the Psychological Strength of the Migrant Workers through Technological Aid

COVID-19 pandemic has distorted the psychological set up of the migrant workers. No earnings of migrant workers at present is less painful than the fear of losing the job permanently. Apprehension of finding difficulty to manage the household in days to come is creating suicidal ideation among some of the migrant workers. Amid acute pressure and frustration, some of the migrant workers are vigilant in social media with the hope of getting some chance somewhere to get engaged. To help the migrant workers to come out of psychological and emotional distortions, counseling through technological aid is very much useful. Through information and communication technology, they can be electronically linked to the counselors for appropriate moral suasion. As most of the migrant workers are accessing TV, programs related to issues of migrant workers and probable solutions for those may be broadcast. To upgrade their skill during their stay at home, skill-based programs may be aired through all social media and electronic media with incentives. There should be a provision for the migrant workers to file their grievances electronically and an e-grievance redressal cell must be there for speedy solution. NGOs need to be promoted to help them out electronically. The following steps are necessary for them from this difficult situation.

- They need to be assured that they are not alone; the whole society is with them.
- Every employer needs to assure them that when work will resume, they will be given first preference. They are not going to loose their job.
- The government needs to make counseling through social media that migrants are the real developer of the economy. In any aspect, they will be protected. So they need not panic.
- Migrants need to be provided in every type of information.
- Proper planning will be made so that they can reach their home safely.
- They need to be supplied food, shelter and health facilities on emergency basis.
- Mental health facility needs to be supplied to these people; otherwise, it will create another problem in post COVID situation.

9.6 Conclusion

It is concluded that the psychological status of most of the repatriated workers has deteriorated since the inception of COVID-19 pandemic. Most of them are stressed, getting irritated and frustrated. Knowing the fact that the returnees are causing the spread of the positive cases in the state, every returnee migrant is having the fear that s/he may be the next positive case of the locality. This fear is killing their intent to leave peacefully. As such, they are economically stressed for not having present income or savings. They are mostly living on transfer earnings. Adding to that, psychological stress making their lives miserable. Suggestions prescribed above may help them a lot pathologically. With the help of technological intervention, particularly by using fuzzy logic, the issues encountered by the migrants can be identified and mapped properly for addressing them.

References

1. Choudhuri, R., COVID 19 pandemic: Mental health challenges of internal migrant workers of India, Elsevier Public Health Emergency collection. *Asian J. Psychiatry*, 54, 102254, 1–4, 2020.
2. Singh, O.P., Mental health of migrant laborers in COVID-19 pandemic and lockdown: Challenges ahead. *Indian J. Psychiatry*, 62, 3, 233–234, 2020.
3. Kumar, K. *et al.*, The psychological impact of COVID-19 pandemic and lockdown on the migrant workers: A cross-sectional survey, Elsevier Public Health Emergency collection. *Asian J. Psychiatry*, 53, 102252, 2020.
4. Bhugra, D., Gupta, S., Bhui, K., WPA guidance on mental health and mental health care in migrants. *World Psychiatry*, 10, 2–10, 2011.
5. Tarsitani, L. and Biondi, M., Migration and mental health: New challenges. Riv. Psichiatr., 51, 45–46, 2016.
6. Mazzetti, M., Tarsitani, L., Geraci, S., The role of migration in postconflict recovery, in: *Book of Best Practices: Trauma and the Role of Mental Health in Post-Conflict Recovery*, R.F. Mollica, R. Guerra, R. Bhasin, J. Lavelle (Eds.), Harvard Programme in Refugee Trauma, Boston, MA, USA, 2006.
7. Vasantha Kandasamy, W.B. and Indra, V., Applications of fuzzy Cognitive maps to determine the maximum utility of a route, in: *J. Fuzzy Math.*, vol. 8, pp. 65–67, International Fuzzy Mathematical Institute, Los Angeles, USA, 2000.
8. Papageorgiou, E.I. and Kontogianni, A., Using fuzzy cognitive mapping in environmental decision making and management: A methodological primer and an application, in: *International Perspectives on Global Environmental Change*, S. Young (Ed.), pp. 427–450, InTech, London, UK, 2012.
9. NSSO report 2007-08, Ministry of statistica and programme implementation, Government of India, in: *Census of India Report 2011*, Ministry of Home Affairs, Government of India, 2001 and 1991.

10

Fuzzy-Based Edge AI Approach: Smart Transformation of Healthcare for a Better Tomorrow

B. RaviKrishna[1], Sirisha Potluri[2], J. Rethna Virgil Jeny[1], Guna Sekhar Sajja[3] and Katta Subba Rao[4]

[1]Department of Computer Science and Engineering, Vignan Institute of Technology and Science, Deshmukhi(V), Pochampally(M), Yadadri Bhuvangiri, Telangana, India
[2]Department of Computer Science and Engineering, Faculty of Science and Technology-IcfaiTech, ICFAI Foundation for Higher Education, Hyderabad, India
[3]Information Technology Department, University of the Cumberlands, Williamsburg, United States
[4]Department of Computer Science and Engineering, B V Raju Institute of Technology, Narsapur, Medak (District), Telangana, India

Abstract
Fuzzy-based edge artificial intelligence is a considerably enhanced solution when compared to the conventional cloud-based artificial intelligence models. Its advantage is observed through real-time data analysis for its local data analysis. The proposed model decreases latency as there are no data transfer and allowing for quick, efficient, and real-time decisions. Applications of fuzzy-based edge artificial intelligence are health analytics, autonomous vehicles, self-driving cars, smart traffic management, robots in manufacturing works, video analytics, etc. As edge artificial intelligence does not need a connection, results into a cost, power, and energy effective solution when compared to cloud-based artificial intelligence models. This technique empowers the construction and management of improved healthcare services that are fairly valuable in the healthcare industry. Fuzzy-based edge AI ensures security and privacy, as the utmost of the data gets processed locally due to its on premise availability, whereas cloud-based artificial intelligence models allow data transfer to other locations can open up access to vulnerabilities.

**Corresponding author*: sirisha.vegunta@gmail.com

Sachi Nandan Mohanty, Prasenjit Chatterjee and Bui Thanh Hung (eds.) Fuzzy Computing in Data Science: Applications and Challenges, (181–196) © 2023 Scrivener Publishing LLC

In our research, we outline fuzzy-based edge AI approach for the smart transformation of healthcare for a better tomorrow.

***Keywords*:** Machine learning, healthcare, fuzzy computing, edge computing, artificial intelligence

10.1 Significance of Machine Learning in Healthcare

The healthcare domain is an imperative industry that provides value-based overhaul to masses of people and is one of the vital income sources. Currently, the healthcare domain in the US alone produces an income of more than $1.5 trillion. The US government also inhabits more on healthcare domain per heads as associated to most other established/emerging nations. Excellence, value, and consequence are three catchwords that always escort healthcare and guarantee a lot, and these days, healthcare professionals and participants throughout the globe are observing for state-of-the-art ways to bring on this promise. Intelligent technology-assisted smart healthcare, Internet-linked medical devices, and edge machine intelligence is holding the healthcare system to a great level [1–3].

After playing a significant role in patient management, billing, and records maintenance, today's expertise is permitting healthcare professionals to develop substitute staffing representations, IP capitalization, offer smart healthcare, and falling administrative and supply budgets. The healthcare domain is gradually seeking the need and assistance of machine learning models for diagnosis, treatment, and drug inventory for various diseases. Google, in recent times, established a machine learning system to detect cancerous growths/lumps in mammograms. The research community at Stanford University is developing a deep learning system to detect skin tumors. Machine learning in the healthcare domain aids to examine thousands of diverse data points and recommend products, deliver sensible risk scores, defined resource apportionment, and has several other significant applications. The objective of our research is to discuss various significant applications of ML in healthcare, and in what way they outlook to change the system to imagine the healthcare industry.

The progressively increasing number of substantial submissions of ML in the healthcare industry permits us to preview the future, to serve numerous patients with its incredible data collection, investigation, and innovation work. Machine learning-based applications are embedded with various devices/machines to retrieve real-time patient data and increase the efficiency of novel treatment opportunities, which were not available before [4–8].

10.2 Cloud-Based Artificial Intelligent Secure Models

Cloud is a dynamic and flexible environment that provides and offers services to various users through efficient resource management [9–11]. Cloud-based applications have perceived exponential progression in their ability to consume enormous quantities of data and yield accurate results. AI-enabled cloud-based models are providing efficient healthcare solutions due to the availability of required resources, scalability of the services, cost effective models and readily offered facilities. Data analytics in the healthcare industry is providing significant insights into data. Telemedicine is creating these insights accessible at every point of care for improved, healthier, and more dedicated diagnosis. Both improvements create more reliable services to bridge the gap between patients and healthcare providers [12–18].

10.3 Applications and Usage of Machine Learning in Healthcare

Various applications and usage of machine learning models in healthcare are shown in Figure 10.1.

10.3.1 Detecting Diseases and Diagnosis

One of the dominant machine learning usage and applications in the healthcare industry is the detection and analysis of diseases, infections, and diseases. Various diseases, such as tumors, are rough to detect during the early stages, dangerous genetic diseases and infections caused by pathogens, etc. IBM Watson Genomics is a key illustration of how incorporating perceptive computing with genome structure and cancer sequencing can aid in constructing a fast identification and diagnosis.

10.3.2 Drug Detection and Manufacturing

One of the main scientific usage and applications of machine learning in healthcare is the early-stage drug detection process. Research and advance technologies, such as advanced sequencing and precision medicine, can support in discovering alternative procedures for the remedy

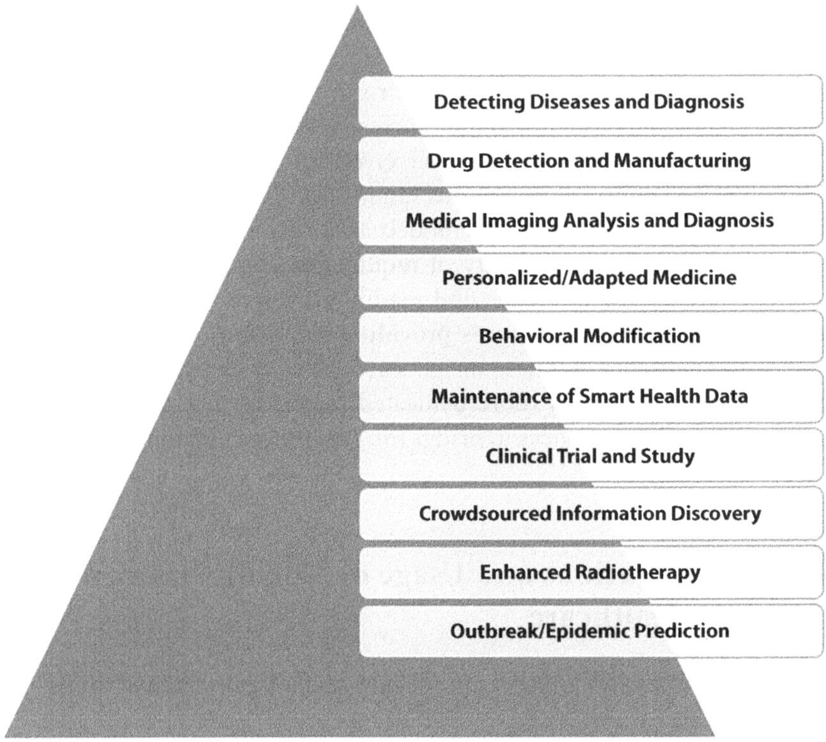

Figure 10.1 Applications and usage of machine learning in healthcare.

of multifactorial illnesses. Machine learning practices implicate unsupervised-based learning, which can recognize outlines in data to proceed toward next-generation exploration and advancements. Microsoft developed Hanover by using machine learning-based technologies for several ingenuity-based AI systems for cancer conduct and personalizing medication amalgamation for severe myeloid leukemia.

10.3.3 Medical Imaging Analysis and Diagnosis

With machine learning, deep learning, and computer vision technologies, Microsoft developed InnerEye for image analytics and everything on image diagnostic tools. Medical image detection deals with the difficulty of detecting a certain portion/element in a medical image. As medical images are volumetric, efficient parsing is observed through AI-driven diagnostic methods.

10.3.4 Personalized/Adapted Medicine

Personalized/adapted conducts can not only be further operative by combining specific health with prophetic analytics but are also suitable are for advanced research and improved disease assessment/treatment. IBM has developed Watson Oncology to understand patient medical history to aid generate several treatment opportunities.

10.3.5 Behavioral Modification

Behavioral modification is a crucial measure of protective medicine, and machine learning-based models are researching the arenas of cancer inhibition and identification that allows us to understand our comatose conduct and make crucial modifications.

10.3.6 Maintenance of Smart Health Data

With the advent of machine learning models such as document/image classification, OCR recognition techniques and smart health records in healthcare a lot of time, energy, and money is saved. Google's Cloud Vision API is using machine learning-based handwriting recognition technology to develop the next generation smart and intelligent, health record maintenance system.

10.3.7 Clinical Trial and Study

Machine learning has numerous potential applications and importance in the area of clinical trials and research. Using ML-based predictive analytics, we can classify prospective clinical trial observations, and it can aid researchers to pull a pool from an extensive variety of data sources, such as preceding doctor appointments, social media, and other related vital information.

10.3.8 Crowdsourced Information Discovery

Crowdsourcing is altogether the indignation in the health field these days, letting researchers and medical practitioners contact a vast volume of information given by individuals based on their interests. Apple's ResearchKit permits users to access collaborative apps, which apply machine learning constructed facial recognition to identify and treat Autism and Parkinson's illness. IBM newly aligned with Medtronic to decipher, collect, and deliver real-time diabetes data founded on the crowdsourced data. With the progressions being made in the Internet of things, the healthcare industry is still identifying new and novel ways to handle tough-to-diagnose scenarios and support the overall progress of analysis, diagnosis, and medication.

10.3.9 Enhanced Radiotherapy

One of the ultimately preferred applications of machine learning in healthcare exists in the field of radiology. Medical image analysis has several distinct variables, which are interpreted with complex equations based on trained machine learning-based models. These models study from the huge training set of different available samples to easily diagnose and find the diseases. Google's DeepMind Health is allowing efficient classification of objects into various categories, such as normal/abnormal, lesion/non-lesion, usual/unusual, etc. to develop efficient machine learning algorithms.

10.3.10 Outbreak/Epidemic Prediction

Artificial intelligence-based technologies are nowadays being set to use in observing and predicting outbreaks/epidemics around the world. Data scientists have access to a huge amount of data gathered from satellites, social media, websites and clinics. Artificial neural networks aid to collate this data, predict and forecast everything from malaria epidemics to SARS-CoV-2. Predicting and forecasting these outbreaks/epidemics is especially useful in third-world nations as they have a deficiency in essential medical infrastructure and informative systems. ProMED-mail is a cloud-based epidemic reporting platform that observes evolving diseases, developing scenarios of epidemics and delivers outbreak intelligence in real time [19–23].

10.4 Edge AI: For Smart Transformation of Healthcare

10.4.1 Role of Edge in Reshaping Healthcare

Edge computing and artificial intelligence promise to create healthcare provision cheaper, well, easier, and improved for everyone. Currently, infectious diseases and epidemics are on the rise. Uncertainties about COVID-19 disease have instigated people to put off regular examinations and treatments, constructing chronic situations worse and guaranteeing

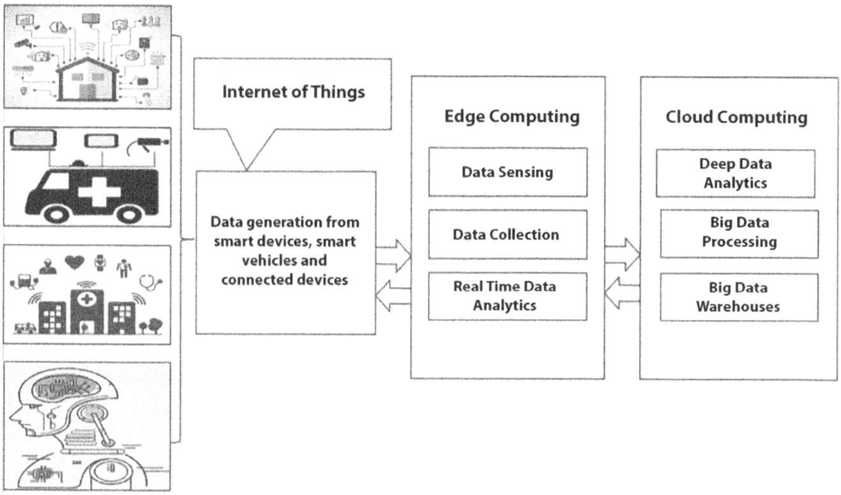

Figure 10.2 Role of edge in healthcare.

an additional 20% rise in additional mortality. According to the report of the World Health Organization, we are now facing a universal deficiency of 7 million healthcare workers and this number could reach more than 13 million by 2040.

To meet several demands and undergo reforming transformation in healthcare, the edge is providing instant solutions as shown in Figure 10.2. Currently, edge computing strategies are being used to control and monitor patients remotely, systematize the automation to deliver healthcare, take along artificial intelligence to advance the speed and precision of diagnoses, track the medicine/vaccine supply chain, etc. That is because edge AI puts data/information/statistics, analytics/decision system, and processing/implementation control where it is most required such as in the ambulance, hospital, operating room, homes, and host bodies.

10.4.2 How AI Powers the Edge

We cannot discourse about edge computing-powered healthcare without also discoursing about artificial intelligence. It is not sufficient to gather data from smart devices, patients, smart vehicles, caregivers, etc., it is also essential to examine it and retort in real time. Progressively, that practice is being accomplished by smart machines using edge technology. According to the survey of IoT security firm Zingbox, the average hospital bed takes 10 to 15 linked devices. Similarly, a survey by Optum—the technology division of UnitedHealth Group tells that 40% of healthcare administrators plan to install AI to study data collected by the IoT devices to monitor the health or diseases or condition of a patient. In more contemporary times, AI has an extent throughout the healthcare industry with its digital landscape. From being on mainframes, workstations and computers, PDAs to the cloud computing and now to the "edge computing." It is this final part that is specifically enthralling. Edge AI is, therefore, well thought-out to be the ensuing chapter in the progression of AI. Conferring to the newest market research testimony of MarketsandMarkets the edge technology will be worth more than $15 billion in the coming 5 years. The prevalent assumption of healthcare IoT devices is the key reason that is motivating its demand. It benefits healthcare administrators to get in trace with their patients in real time [24–28].

10.5 Edge AI-Modernizing Human Machine Interface

Edge computing is serving to reshape the present healthcare industry. As specified above, there are several benefits of Edge AI and is proven as the best option for both the patients and the healthcare providers.

10.5.1 Rural Medicine

Providing value healthcare to secluded rural areas has been a great challenge in past. Even today, with discoveries in telemedicine and additional easily accessible health data, medical providers have resisted providing fast, excellent attention to people who live far from hospitals and observe inadequate net access. Normal healthcare databases challenge substantial experiments here along with their connectivity difficulties. But the edge computing applications overcome all these difficulties with their novel approaches. IoT-based healthcare devices can extend the grasp of existing networks, permitting the medical workforce to contact critical patient data even in areas with insufficient connectivity. This use case on Edge AI, showing its capacity to significantly spread the best of health services.

10.5.2 Autonomous Monitoring of Hospital Rooms—A Case Study

One of the USPs of AI, in all-purpose, is prospective to automate things. AI processes can gather data using a congregation of sensors and investigate it to retort most suitably. Edge AI proceeds this up a notch. Computer vision, data analytics, and artificial intelligence models enable autonomous observing of hospital rooms, infrastructure, and patients. Take for example fall discovery. Many IoT-enabled wearable devices these days arise with the capacity to sense if a person falls unexpectedly with specific hardware. Processes of edge AI in these wearable devices can be competent to sense falls in an immediate and even communicate to caregivers. In maximum cases, this is treated as a life-saving option. Ex: Fall detection option available on Apple Watch.

Edge AI can also be of immense support in the observing of important signs. Medical devices intended to record significant data like heart rate, body temperature, breathing rate, blood sugar, blood pressure, hypertension, etc. can influence AI to notice any deviation in an instant. These devices could then alert the hospital staff/caregivers and they can take necessary action. For the patient, this is not only precarious but also progresses the overall treatment and medication [29–31].

10.6 Significance of Fuzzy in Healthcare

10.6.1 Fuzzy Logic—Outline

Nevertheless, the enormous improvement and increasing the elderly people implies more resources for aftercare, paramedical care and natural assistance in their habits. The fuzzy logic was introduced with the Lotfi Zadeh's proposal of fuzzy in 1965. The term fuzzy refers to the things that are not clear and or are vague, which provides very valuable flexibility for reasoning, that resembles human reasoning in many ways, and hence we can consider uncertainties and inaccuracies of any situation. The term Crisp value means a precise can be referred to a Boolean logic system. In fuzzy logic, there are intermediate values present that are partially true and partially false. Fuzzy sets can be considered as an extension of classical sets. It swallows partial membership, which means that it contains elements that have varying degrees of membership in the set. Classical set contains elements that satisfy precise properties of membership which fuzzy set contains elements that satisfy imprecise properties of membership.

A fuzzy set $\tilde{A}$ in the universe U can be defined as a set of ordered pairs and it can be represented mathematically as

$$\tilde{A} = \{(y, \mu_{\tilde{A}}(y)) / y \in U\}$$

$\mu_{\tilde{A}}(y) =$ Membership of Y in $\tilde{A}$ and $\mu_{\tilde{A}}(y) \in [0,1]$.

The way of representing a Fuzzy set as:

$$\tilde{A} = \left\{ \frac{\mu_{\tilde{A}}(y_1)}{y_1}, \frac{\mu_{\tilde{A}}(y_2)}{y_2}, \frac{\mu_{\tilde{A}}(y_3)}{y_3}, \ldots \right\} = \sum_{i=1}^{n} \frac{\mu_{\tilde{A}}(y_i)}{y_i} \quad \text{or}$$

$$\tilde{A} = \left\{ \left(y_1, \mu_{\tilde{A}}(y_1) \right), \left(y_2, \mu_{\tilde{A}}(y_2) \right), \left(y_3, \mu_{\tilde{A}}(y_3) \right) \right\}$$

$\mu_{\tilde{A}}(y)$ is the contribution or importance of y. Fuzzy logic being a computing approach on "degree of truth," rather than conventional Boolean logic classification "true or false" (1 or 0) the one modern computing requirement is needed. The applications of fuzzy logic are numerous. In real world scenarios like weather forecasting business decision making, an impeccable role in healthcare support systems and medicinal making artificial intelligence and neural network which are in trend today, fuzzy logic can be used vibrantly used in all these fields and gaining more popularity. The reason why fuzzy logic applies so well in real life is because in most of the scenarios, we cannot have a distinct value, rather based on an approximation. For example, the no. of red blood cells in your blood can not be actually counted but we can have an estimated value, or the growth of bacteria in a particular environment can only be predicted, but cannot be counted and have the correct value as well. Fuzzy logic works on the logic based on Membership and degree of membership which ranges from 0 to 1. Depending on the membership, function depends on different membership values ranging from 0 to 1, which will decide the impact of that particular characteristic nature or behavior of a member.

10.6.2 Fuzzy Logic-Based Smart Healthcare

A single disease may be symptomatic in different forms depending on the patient's health conditions and inheritance. In the same way, a single symptom may correspond to different diseases. This description of diseases entitles in linguistic terms that are not exact and may be vague. To handle this inaccuracy and uncertainty, we need an efficient mechanism. Medical diagnosis is fundamentally a pattern grouping phenomenon. Depending on some input delivered by a patient, a professional provides inference based on its information and knowledge, which is usually warehoused in a binary form, and lastly, thc outcome is considered, i.e., either the infected person is suffering from a definite disease or not. There are various properties and facilities in fuzzy set theory, and due to which, it is made suitable for medical diagnosis. The result and discovery from the study have revealed that the artificial intelligence-based technique with fuzzy logic can add a reliable outcome to advise the disease [32–34].

10.6.3 Medical Diagnosis Using Fuzzy Logic for Decision Support Systems

In this chapter, medical diagnosis support system will be discussed which is generated using fuzzy logic that focuses on the medical support system. With the advent of fuzzy logic to the model physician diagnosis decisions, it will have much more accuracy than the conventional decisions that are being considered earlier. Some of the fuzzy algorithmic approaches for the decision support systems are multicriteria decision methods and aggregation approaches using the fuzzy approach.

First of all, we should consider a basic normative model for the decision theory: for the set of alternatives with feasible options, X, the set of states relevant to the context of the experiment, the resulting scenarios E, and utility function u that order the spaces of events with respect the desirability. The highest yields of utility will lead to the decision concerning uncertainty and possibility. The feasible alternatives may be explicitly defined with implicit enumerable constraints. The developing model can be extended to work on different objectives. The tradeoff is the decision making point toward the other objective. The real-world scenarios may have the add-on vague semistructured situations that may arise like:

- ✓ The alternative cannot be determined crisp, though it can be gained to a certain degree of membership,
- ✓ The situation may not be represented mathematically, but can be explained and represented by vague ideas and linguistic terminologies.
- ✓ The utility depends on several categorical decisions and risk comportment
- ✓ The functional dependencies are unable to depict.

We restrict our discussion to one major facility of a multiattribute decision model (MADAM). In turn, to provide the comprehension of the fuzzy model's area, let us understand the possible degree of desirability. The aggregation approach will be usually consisting of two phases.

✓ Phase 1: It is of the judgment among all goals and decision alternatives.
✓ Phase 2: It is the ranking of the decision alternatives to the summarized judgments.

A multiattribute decision-making system usually consists of a set of alternatives. These alternatives will be treated as variables and will be compared with respect to multiple, and using conflict attributes.

The aim to determine the ranking with respect to all the attributes in a given scenario is based on classical utility theory and performance with respect to each attribute and its importance against the objectives. The problem must be assessed before aggregating the subjective assessments together to determine the overall performance of each among all attributes in the situation. The overview of these processes is shown in Figure 10.3. After identifying the alternatives and attributes for the multiattribute decision making model, the pairwise comparison and assessment of the performance of those attributes will be computed and then the computation of alternative performance using fuzzy expert analysis will take place. Along with that we also compute the alternative importance of the fuzzy expert analysis. The consecutive processes will take place to the final

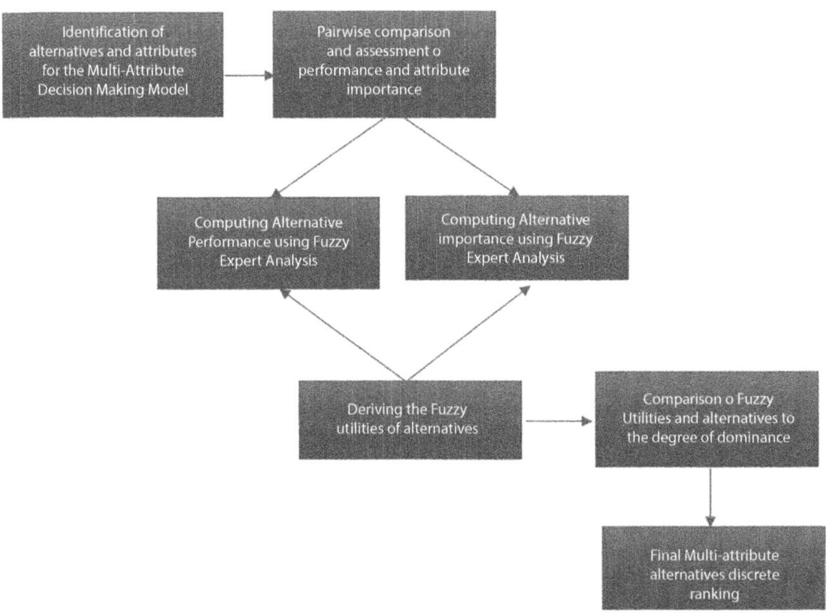

Figure 10.3 The approach of multiattribute decision model.

multiattribute alternatives discrete ranking after deriving the fuzzy utilities of alternatives. In this process, the degree of dominance of the utilities and alternatives will be used for decision making for computing the raking.

10.6.4 Applications of Fuzzy Logic in Healthcare

The complicated medical procedures result from the lack of proper and accurate information, imprecise information and contradicting nature raising the advanced developments in diagnostic procedures and treatments. To address these issues, fuzzy logic can alone be used in a hybrid manner with other methods like neuro-fuzzy and fuzzy-NN and fuzzy Bayesian applications. It has a wide scope of applications in Medical Research. Ranking of risk factors and test results and their performance is important for decision making process. Clustering is used for the split of data (referred to as clusters) to derive the hidden patterns for the objects with similar properties and different behaviors. Medical Image processing has a vital role in medical decision making. 2-Dim and d3-Dim images are the main sources of magnetic resonance image processing, digital mammography and positron, emission tomography and topography tests. The images will generally be textures with noise and the acquisition of inaccuracies in the expected portions. In these areas, using the clustering techniques is more useful to make the precise decision with imprecise conditions and vagueness [35–37].

10.7 Conclusion and Discussions

The requirement for real-time, reasonable, and effective smart healthcare facilities is increasing rapidly due to the scientific revolution and surge of population. To satisfy the growing demands on this precarious infrastructure, there is a necessity for intelligent approaches to survive the prevailing obstacles in the healthcare area. In this esteem, fuzzy-enabled edge computing expertise can decrease latency, communication cost, and energy consumption by keeping processes nearer to the data sources in contrast to the typical centralized cloud-enabled healthcare systems. Artificial intelligence delivers the possibility of perceiving and forecasting high-risk infections in advance, declining medical charges for patients, and proposing proficient treatments. The main objective of this research is to highlight the assistances of the agreement of fuzzy-based edge intelligent technology with AI in smart healthcare systems.

References

1. Yu, K.H., Beam, A.L., Kohane, I.S., Artificial intelligence in healthcare. *Nat. Biomed. Eng.*, 2, 719–731, 2018.
2. Jiang, F., Jiang, Y., Zhi, H. *et al.*, Artificial intelligence in healthcare: Past, present and future. *Stroke Vasc. Neurol.*, 2, 4, 230–243, 2017.
3. Davenport, T. and Kalakota, R., The potential for artificial intelligence in healthcare. *Future Healthc. J.*, 6, 2, 94–98, 2019.
4. Amin, S.U. and Hossain, M.S., Edge intelligence and internet of things in healthcare: A survey. *IEEE Access*, 9, 45–59, 2021.
5. Keshavarzi, A. and van den Hoek, W., Edge intelligence—On the challenging road to a trillion smart connected IoT devices. *IEEE Des.Test*, 36, 2, 41–64, 2019.
6. Liu, Y., Peng, M., Shou, G., Chen, Y., Chen, S., Toward edge intelligence: Multiaccess edge computing for 5G and internet of things. *IEEE Internet Things J.*, 7, 8, 6722–6747, 2020.
7. Du, Y., Wang, Z., Leung, V.C.M., Blockchain-enabled edge intelligence for IoT: Background, emerging trends and open issues. *Future Internet*, 13, 248, 1–21, 2021.
8. Rausch, T. and Dustdar, S., Edge intelligence: The convergence of humans, things, and AI. *2019 IEEE International Conference on Cloud Engineering (IC2E)*, pp. 86–96, 2019.
9. Satpathy, S., Mangla, M., Mohanty, S.N., Potluri, S., GA-based iterative optimization system to supervise adaptive workflows in cloud environment, in: *Emerging Technologies in Data Mining and Information Security. Advances in Intelligent Systems and Computing*, vol. 1300, A.E. Hassanien, S. Bhattacharyya, S. Chakrabati, A. Bhattacharya, S. Dutta (Eds.), Springer, Singapore, 2021.
10. Potluri, S., Mangla, M., Satpathy, S., Mohanty, S.N., Detection and prevention mechanisms for DDoS attack in cloud computing environment. *2020 11th International Conference on Computing, Communication and Networking Technologies (ICCCNT)*, pp. 1–6, 2020.
11. Malladi, A. and Potluri, S., A Study on technologies in cloud-based design and manufacturing. *Int. J. Mech. Prod. Eng. Res. Dev.*, 8, 187–192, 2018.
12. Gunturi, M., Kotha, H., Potluri, S., An Iot based solution for health monitoring using a body-worn sensor enabled device. *J. Adv. Res. Dyn. Control Syst.*, 10, 646–651, 2018.
13. Potluri, S. and Avinash, M., Health record data analysis using wireless wearable technology device. *J. Adv. Res. Dyn. Control Syst. (JARDCS)*, 10, 9, 696–701, 2018.
14. Ravikrishna, B. and Singh, H.P., Big data classification using hybrid parallel Lfr-Cm at cloud environment. *IJAST*, 29, 08, 6469, May 2020.

15. Potluri, S., Sunaina, S., Neha, P., Govind, C., Raghavender, J., Gupta, V.M., A secure cloud infrastructure towards smart healthcare: IoT based health monitoring. *Cloud Secur. Technique. Appl.*, 63, 63–82, 2021.
16. Mohanty, S.N., Mohammad, G.B., Potluri, S., Ramya, P., Lavanya, P., Next generation cloud security: State of the art machine learning model. *Cloud Secur. Technique. Appl.*, 125, 125–144, 2021.
17. Potluri, S., Sarkar, A., Yasin, E.T., Mohanty, S.N., IoT enabled cloud based healthcare system using fog computing: A case study. *J. Crit. Rev.*, 7, 6, 1068–1072, 2020.
18. Mohanty, S.N., Potluri, S., Prakash, V.B., Srinath, B., Manjunath, B., Cloud security concepts, threats and solutions: Artificial intelligence based Approach. *Cloud Secur. Technique. Appl.*, 1, 1–20, 2021.
19. Bhardwaj, R., Nambiar, A.R., Dutta, D., A study of machine learning in healthcare. *2017 IEEE 41st Annual Computer Software and Applications Conference (COMPSAC)*, pp. 236–241, 2017.
20. Callahan, A. and Shah, N.H., Chapter 19 - Machine learning in healthcare, in: *Advances in Clinical Informatics*, A. Sheikh, K.M. Cresswell, Wright, D.W. Bates (Eds.), pp. 279–291, Academic Press, Canada, 2017.
21. Siddique, S. and Chow, J.C.L., Machine learning in healthcare communication. *Encyclopedia*, 1, 1, 220–239, 2021.
22. Waring, J., Lindvall, C., Umeton, R., Automated machine learning: Review of the state-of-the-art and opportunities for healthcare. *Artif. Intell. Med.*, 104, 1–12, 2020.
23. de la Torre, J., Marin, J., Ilarri, S., Marin, J.J., Applying machine learning for healthcare: A case study on cervical pain assessment with motion capture. *Appl. Sci.*, 10, 17, 1–28, 5942, 2020.
24. Ray, P.P., Dash, D., De, D., Edge computing for Internet of Things: A survey, e-healthcare case study and future direction. *J. Netw. Comput. Appl.*, 140, 1–22, 2019.
25. Li, X., Huang, X., Li, C., Yu, R., Shu, L., EdgeCare: Leveraging edge computing for collaborative data management in mobile healthcare systems. *IEEE Access*, 7, 22011–22025, 2019.
26. Pustokhina, I.V., Pustokhin, D.A., Gupta, D., Khanna, A., Shankar, K., Nguyen, G.N., An effective training scheme for deep neural network in edge computing enabled internet of medical things (IoMT) systems. *IEEE Access*, 8, 107112–107123, 2020.
27. Rahman, M.A. and Hossain, M.S., An internet of medical things-enabled edge computing framework for tackling COVID-19. *IEEE Internet Things J.*, 8, 21, 15847–15854, 2021.
28. Klonoff, D.C., Fog Computing and Edge Computing architectures for processing data from diabetes devices connected to the medical Internet of Things. *J. Diabetes Sci. Technol.*, 11, 4, 647–652, 2017.
29. Hassan, N., Gillani, S., Ahmed, E., Yaqoob, I., Imran, M., The role of edge computing in internet of things. *IEEE Commun. Mag.*, 56, 110–115, 11, 2018.

30. Alshehri, F. and Muhammad, G., A comprehensive survey of the internet of things (IoT) and AI-based smart healthcare. *IEEE Access*, 9, 3660–3678, 2021.
31. Sun, L., Jiang, X., Ren, H., Guo, Y., Edge-cloud computing and artificial intelligence in internet of medical things: Architecture, technology and application. *IEEE Access,* 8, 101079–101092, 2020.
32. Rizwan, A., Zoha, A., Mabrouk, I.B., Sabbour, H.M., Al-Sumaiti, A.S., Alomainy, A., Imran, M.A., Abbasi, Q.H., A review on the state of the art in atrial fibrillation detection enabled by machine learning. *IEEE Rev. Biomed. Eng.,* 14, 219–239, 2021.
33. Sellam, V., Kannan, N., Basha, H.A., An effective fuzzy logic based clustering scheme for edge-computing based internet of medical things systems, in: *Cognitive Internet of Medical Things for Smart Healthcare. Studies in Systems, Decision and Control,* vol. 311, A.E. Hassanien, A. Khamparia, D. Gupta, K. Shankar, A. Slowik (Eds.), Springer, Cham, 2021.
34. Kakkar, A. and Farshori, A., Collaborative Medical Inventory Resources Using Edge Computing – A Solution to Serve Critical Healthcare Requirements at Public Hospitals. *Amity International Conference on Artificial Intelligence (AICAI)*, pp. 647–652, 2019.
35. Mulimani, M.S. and Rachh, R.R., Edge computing in healthcare systems, in: *Deep Learning and Edge Computing Solutions for High Performance Computing. EAI/Springer Innovations in Communication and Computing,* A. Suresh and S. Paiva (Eds.), Springer, Cham, 2021.
36. Tsai, Y.T. and Lin, Z.Y., "*A Survey Edge Computing Bioinf. Health Informatics**," *2020 IEEE Int. Conf. Bioinf. Biomedicine (BIBM)*, 2020, pp. 2203-2208, 2020.
37. Kumar, M.G.S., Dhulipala, V.R.S., Baskar, S., Fuzzy unordered rule induction algorithm based classification for reliable communication using wearable computing devices in healthcare. *J. Ambient Intell. Hum. Comput.,* 12, 3515–3526, Mar 2021.

11

Video Conferencing (VC) Software Selection Using Fuzzy TOPSIS

Rekha Gupta

Lal Bahadur Shastri Institute of Management, Delhi, India

Abstract

The coronavirus-19 pandemic had bought drastic changes to the teaching and learning environments. With learning setups shifting entirely to the online medium, the higher education institutions were faced with the dilemma of selecting the video-conferencing software along with technical infrastructure setup, users training issues. The selection decisions were frequently done randomly with little IT expertise to guide. Even soon, the hybrid mode (physical teaching combined with online mode) would be the teaching mediums offered by the higher education institutions as for now they have some familiarity in terms of usage and technical setup of the software. Several advantages, as well as disadvantages of online teaching have been realized and experienced during the said times. Further, the institutions faced the dilemma of not selecting the most viable video conferencing solution as per their requirements. Therefore, the author attempts to provide a comprehensive guide for selecting the most apt video-conferencing software for the higher education institution based on their requirements and further attempts to rank the available options using Fuzzy-TOPSIS method.

Keywords: Online teaching, higher education institutions, hybrid mode of teaching, videoconferencing software, Fuzzy TOPSIS

11.1 Introduction

Novel coronavirus virtually impacted all sectors worldwide and had been instrumental in transforming education sector enormously with

Email: rekhagupta@lbsim.ac.in

Sachi Nandan Mohanty, Prasenjit Chatterjee and Bui Thanh Hung (eds.) Fuzzy Computing in Data Science: Applications and Challenges, (197–214) © 2023 Scrivener Publishing LLC

millions of teachers and students shifting from traditional mode to online mode. UNESCO (2020) also called this as an education emergency [1]. International Labor organization (ILO), 2020 reported that disruption in education sector had impacted several million schools, educational institutes, teachers, and students [2].

Most higher education institutions (HEI) struggled overnight to select, install the basic video conferencing software to impart the teaching lessons to have unhindered learning process. The software for online teaching was mostly selected randomly/recommended/suggested by the little IT team available with the institutions. Most teaching institutions, organizations, teachers, and students were completely new to the interface of the video conferencing software and hence, struggles were apparent in the initial days with all stakeholders. With or without training, slowly the stakeholders got adapted to the usage and features of the videoconferencing software.

This scenario also provided an ample market opportunity for the companies selling the videoconferencing software for educational institutions. Some of the prominent companies in the videoconferencing market were Microsoft with MS Teams, Cisco with Cisco WebEx, Google with Google Meet, Lark, Zoom, etc. Few other companies, such as TCS-ion sold their Learning Management System bundled with video-conferencing software. Prior to the pandemic, the companies had been offering this software as part of the bundle package along with mail subscription and office suite products. The pandemic bought ample opportunities for companies to view this product in huge demand and as a revenue generator. Subsequently, many of them utilized the opportunity to invest, enhance features, commercialize, and market it properly.

Online teaching has been extensively reviewed in research studies for its advantages, as well as its disadvantages. Some of the notable advantages listed are anytime, anyplace and from anyone kind of teaching. It is more affordable and saves time for producing measurable results. Other advantages were repeated access to recorded lectures, flexibility in increasing number of students, providing learning opportunities to distant students. Further, online teaching ensures consistency, convenience, and same experiential learning to near and distant students.

Few disadvantages enlisted were the dependency of administrators, teachers, and students on information and communication technologies, requirement of continual technical training and support, monitoring of online teaching quality, rigor of assessment and learning and teaching motivation were other factors that were recorded as degrading factors of online education. Learner's isolation, frustration, and lack of peer group learning were also recorded as other disadvantages.

With the pandemic stage set in 2020, it became imperative that online teaching was the only available via medium to continue with the teaching, learning process. Hence, the immediate requirement was to select the video conferencing software for delivering the same. As institutions grappled with the choices available to them, it was felt to discuss the factors that lead to the selection of a video conferencing software and present a selection mechanism. It is also evident that during the pandemic and post-pandemic times, the hybrid mode of teaching is to stay as institutions have experienced its advantages with stakeholders willing to adapt to the digital mode of learning and teaching.

Therefore, this paper attempts to cover the following objectives:

i. To enlist the major features of Video-conferencing software.
ii. To illustrate the Fuzzy-TOPSIS methodology for video-conferencing selection.
iii. Numerical illustration for selection using real time data from higher education institutes in Delhi-NCR.

11.2 Video Conferencing Software and Its Major Features

11.2.1 Video Conferencing/Meeting Software (VC/MS) for Higher Education Institutes

The VC software provided for meeting online for conduct of regardless of location, role, experience level, language, and device preference, which was of prima face importance in times of coronavirus. The author had tried to scan major videoconferencing software used by Higher-Education institutions particularly the B-schools imparting management and technology education in Delhi-NCR region. The discussion, however, is restricted to common features that are available in all VC software. Thus, an attempt was also made to enlist the features that the institutions considered/would like to consider in near future in case they would want to change the software. The major critical features of VC software are presented below:

i) Meeting Capacity
The number of participants that are allowed in a meeting in the VC software generally varied from 100 to 500. These were live interactions and allowed participants to unmute, speak and open their videos. Higher education institutes generally have a class size of 60 and have 3-6 sections in a batch. Hence, this meeting size was apt for taking classes and batch wise interaction. However, this was insufficient for addressing all the students of the institution together. The alternative option available is a webinar/livestream mode with live feed in which attendees as passive audience with chat option for interaction. The webinar mode could support up to 10,000 audiences. The batch interaction was facilitated through this option.

ii) Meeting Scheduling
The meetings could be prescheduled from the calendar or could be instantly initiated via a call. Some software allowed for web calls as well. The upgraded plans allowed for HD video and audio calling, as well. The calendar also had provisions for showing meeting conflicts, if any. With unlimited meeting call durations, meetings could remain valid for few days to 2 months. The meetings as per requirement could be scheduled as one-time or as recurring. The meeting links notification could be sent to individuals or group with the meeting notes. The meeting reminder was also an inherent feature. Meeting's provision was supported through desktop, browser or mobile depending on the resource's availability. The recipients had provision of accepting, denying, or replying to the meeting notification. The scheduler/presenter had access controls by options of wait/allow to join the meeting. He could also remove an attendee from the meeting. The meeting attendance feature was subsequently added by many vendors. These features allowed teachers to schedule classes, ensure punctuality, attendance, and discipline in classes.

iii) Meeting Control
The presenter had options of audio/video and chat controls. The presenter could disable attendees to mute others or remove from meeting. He could also mute all attendees and limit presentation rights. The presenter could ask the attendees to open their videos. Most VC software allowed

for 25-49 screens to be viewed simultaneously. The provision of pinning a screen was also available for closer viewing and monitoring. This allowed the teacher to monitor the student activities online. This feature was also used by institutions to proctor the online exams. Although, this mechanism had its own limitations, it did make some psychological pressure on the examinees.

iv) Meeting Interaction:
This was facilitated through audio, video, and chat options primarily. The meetings provided for an audio/video preview to prepare oneself for the onscreen presence. The participants had the provision of using natural/customized backgrounds while sharing their on-screen presence. They had the raise hand feature to showcase their queries. Some VC software also looking to provide backgrounds with a running video. The meeting software also automatically adjusted layouts on screen to display most active participants. The presenter window could be zoomed/toggled for having an enlarged and clear view of shared content. Most vendors provided whiteboard feature that simulated a classroom blackboard. The presenters were allowed to share screen with screen annotation facility as well. The provision of chat was used for responding to queries of attendees and providing feedback. Chats could be broadcasted or be directed at individuals or a specific set of individuals. The chats also have emoji support for personalized responses and reducing notification. Chats could be pinned, and specific chat threads could be saved. Announcements, deadlines, or anything with a future reference in a chat could be also pinned to serve as reminders. Some VC software also provided auditorium view of presenter and attendees simulating the sitting in a class/conference. Also, the transcript facilities were provided in the VC software with translation provision in multiple languages for better understanding. Further integration with other applications, such as mail, document etc., could also be managed via the VC software for quick access, composure, and management. Further, enhanced, and improved features included cancellation of background noise, light adjustments, refreshed user interfaces and enriched reliability features along with data saver features on video calls.

v) Individual and Group workspace
The meetings provided a forum for simulating a classroom with all students as attendees and teacher as presenter. The meeting further has breakout rooms' provision for group activities, which is an essential feature in higher education. Further, VC software provided for individual and collaborative group space. The presenter/teacher was assigned exclusive space for content creation in terms of uploading files, assignments, and exercises. This space was viewable and noneditable for the attendees. Further, Individual workspace and group workspace allowed for attendees, in terms of uploading files for assignments and assessments. Further, cloud storage space was also allocated to all users by the meeting software. Almost all VC software allowed for high-speed upload and download of files in multiple file formats and that too with a secure cross platform synchronization.

vi) Assessment Provision
Assessment provision in the meetings was incorporated through polling and quick surveys. Further, a few VC software gave provision for designing a quiz or form like assessments with both multiple choice and subjective assessments. Automatic correction was also supported. Provision of scheduling, distributing and assessment was also inbuilt. Further, the attendees could see their scores and feedback too. Report generation for the given assessment was also included. The test link could be provided in chat or could be entered through a QR code. This was of immense relief to teachers as it provided them with some mechanism for administrating assessments online during pandemic times.

vii) Security
This was a prima face concern as teachers and students have known to be operating in a secure and protected physical boundaries away from online nuisances of auto play of advertisements, security breaches and privacy invasion. One of the important market strategies was to play the security card. Most of VC vendors used anti-hijacking measures and secure controls for anti-abuse measures by using secure service layer (SSL) or transport layer security (TSL) protocols. They attempted to provide

a secure environment for their meetings and messages were encrypted in transit. Further, safety features were continually enhanced from the vendor side by continually verifying and updating their security, privacy, and verification compliance.

viii) Cost

When the pandemic started, most of the available VC softwares were free for education except a few. Others had upgraded business plans that included enlarged meeting capacity with enhanced features. While the free ones required a few compliances from the institutions side for the basic licensing plans. As pandemic progressed, the vendors switched the free plan to business plans for education institutions with only a few notable exceptions. The paid plans and features varied as per different subscription plans.

ix) Infrastructure requirement, Support and Training

Considered essential in the educational sector as all stakeholders were new to the usage of software, it was considered not only important but vital. The training support provided by the vendors tried to reduce the anxiety level of the users. Many big players provided free trainings and had ongoing training schedules. Further, customized training schedules were also provided on request. Most of the vendors provided business hours support for free plans and 24*7 phone and web support for pricier plans.

11.3 Fuzzy TOPSIS

11.3.1 Extension of TOPSIS Algorithm: Fuzzy TOPSIS

Hwang & Yoon [3] proposed the TOPSIS method in 1981. The linguistic TOPSIS method for group decision-making and the choice process was proposed by [4]. According to this technique, the best alternative would be the one that is nearest to the positive ideal solution and farthest from the negative ideal solution [5]. The positive ideal solution is a solution that maximizes the benefit criteria and minimizes the cost criteria, whereas the negative ideal solution maximizes the cost criteria and minimizes the benefit criteria [6–8]. In other words, the positive ideal solution is composed of all best values attainable of criteria, whereas the negative ideal solution consists of all worst values attainable of criteria [9, 10].

The detailed steps are described below:

i) Fuzzy Ratings for Criteria
Identify the ERP vendors and the experts' ratings. Use linguistic rating variable to assess the importance of the criteria as given in Table 11.1.

Table 11.1 Fuzzy linguistic ratings.

Linguistic scale	Corresponding fuzzy numbers
Very High (VH)	(0.9, 1.0, 1.0)
Medium High (MH)	(0.7, 0.9, 1.0)
High (H)	(0.5, 0.7, 0.9)
Medium (M)	(0.3, 0.5, 0.7)
Medium Low (ML)	(0.1, 0.3, 0.5)
Low (L)	(0, 0.1, 0.3)
Very Low (VL)	(0, 0, 0.1)

ii) Fuzzy Ratings for Alternatives
The decision makers use the linguistic rating variable to evaluate the rating of alternatives with respect to each criterion. The linguistic scale used is given in Table 11.2.

Table 11.2 Linguistic scale for alternative rankings.

Linguistic scale	Corresponding fuzzy numbers
Very Good (VG)	(9, 10, 10)
Good (G)	(7, 9, 10)
Medium Good (MG)	(5, 7, 9)
Fair (F)	(3, 5, 7)
Medium Poor (MP)	(1, .3, 5)
Poor (P)	(0, 1, 3)
Very Poor (VP)	(0, 0, 1)

iii) Fuzzy Decision Matrix and Fuzzy Weights
Convert the linguistic evaluation into triangular fuzzy numbers to construct the fuzzy decision matrix and determine the fuzzy weight of each CSF.
The importance is computed as

$$\tilde{x}_{ij} = \frac{1}{K}\left[\tilde{x}_{ij}^{1}(+)\tilde{x}_{ij}^{2}(+)...(+)\tilde{x}_{ij}^{K}\right] \tag{i}$$

$$\tilde{w}_{j} = \frac{1}{K}\left[\tilde{w}_{j}^{1}(+)\tilde{w}_{j}^{2}(+)...(+)w_{j}^{K}\right] \tag{ii}$$

Where $\tilde{x}_{ij}^{K}$ and $\tilde{w}_{ij}^{K}$ are the ratings and the importance weight of the *Kth* decision maker.

iv) Normalized Decision Matrix
Construct the normalized fuzzy decision matrix $\tilde{R} = \left[\tilde{r}_{ij}\right]_{m\times n}$ where *B* and *C* are the set of benefit criteria and cost criteria, respectively and

$$\tilde{r}_{ij} = \left(\frac{a_{ij}}{c_j^*}, \frac{b_{ij}}{c_j^*}, \frac{c_{ij}}{c_j^*}\right), \quad j \in B; \tag{iii}$$

$$\tilde{r}_{ij} = \left(\frac{a_j^-}{c_{ij}^*}, \frac{b_j^-}{c_{ij}^*}, \frac{c_j^-}{c_{ij}^*}\right), \quad j \in C; \tag{iv}$$

$$c_j^* = \max_i c_{ij} \ \text{ if } j \in B;$$

$$a_j^- = \min_i \ a_{ij} \ \text{ if } j \in C;$$

The normalization process ensures that the values of fuzzy numbers lie within range [0, 1].

v) Weighted Normalized Decision matrix
Construct the weighted normalized decision matrix using eq. v

$$\tilde{V} = \left[\tilde{v}_{ij}\right]_{m\times n} i = 1,2,...,m \; j = 1,2,...,n, \tag{v}$$

where $\tilde{v}_{ij} = \tilde{r}_{ij}(.)\tilde{w}_j$

vi) FPIS and FNIS

Determine the Fuzzy Positive Ideal Solution (FPIS, A^*) and Fuzzy Negative Ideal Solution (FNIS, A^-).

$$A^* = \left(\tilde{v}_1^*, \tilde{v}_2^*, \ldots\ldots \tilde{v}_n^*\right) \tag{vi}$$

$$A^* = \left(\tilde{v}_1^-, \tilde{v}_2^-, \ldots\ldots \tilde{v}_n^-\right) \tag{vii}$$

where $\tilde{v}_j^* = (1,1,1)$ and $\tilde{v}_j^- = (0,0,0)$ $j = 1,2,\ldots,n$

vii) Distance Calculation

Calculate the distance of each candidate from FPIS and FNIS.

The distance of each alternative from $A*$ and A^- can be currently calculated as

$$d_i^* = \sum_{j=1}^{n} d\left(\left(\tilde{v}_{ij}, \tilde{v}_j^*\right)\right), i = 1,2,\ldots,m \tag{viii}$$

$$d_i^- = \sum_{j=1}^{n} d\left(\left(\tilde{v}_{ij}, \tilde{v}_j^{--}\right)\right), i = 1,2,\ldots,m \tag{ix}$$

where $d(.,.)$ is the distance measurement between two fuzzy numbers?

viii) Closeness Coefficient

The closeness coefficient is defined to determine the ranking order of all alternatives once their FPIS and FNIS is calculated.

The closeness coefficient of each alternative is calculated as

$$CC_i = \frac{d_i^-}{d_i^* + d_i^-}, \qquad i = 1,2,\ldots,m \tag{x}$$

Candidates are ranked based on their closeness coefficient. The alternative that is closer to FNIS ($A*$) FPIS and farther from FNIS (A^-) is selected. The one with approaching to 1 is selected as the best among feasible alternatives.

11.4 Sample Numerical Illustration

We hereby consider available three products P1, P2 and P3 for selection consideration. All the nine critical factors were evaluated firstly standalone by the experts for importance rating and later relatively to the products considered. A total of 10 IT experts who were instrumental in successful implementation of the VC software at their premises were considered for importance assessment. Further the criteria were classified as cost criteria that were to be minimized and as benefit criteria that were to be maximized. The author had considered cost that includes training cost as the cost criteria with all others being benefit criteria. Table 11.3 presents the criterion rating for each CSF as given by the ten experts. The criteria wise ranking of each product alternative as given by the experts is illustrated in Table 11.4. The Fuzzy Decision Matrix and Fuzzy Weights of Alternatives are recorded in Table 11.5. Table 11.6 records the Fuzzy Normalized Decision Matrix whereas Table 11.7 shows the Fuzzy Normalized weighted decision matrix.

Table 11.3 Importance weight of each criterion.

Criteria	E1	E2	E3	E4	E5	E6	E7	E8	E9	E10
C1	VH	H	MH	H	VH	H	VH	VH	MH	H
C2	L	VL	ML	VL	L	ML	VL	L	VL	ML
C3	H	H	H	VH	H	VH	H	H	VH	VH
C4	VH	VH	H	H	H	VH	VH	VH	MH	VH
C5	M	ML	M	H	H	M	M	M	M	H
C6	H	VH	VH	MH	VH	H	MH	H	VH	H
C7	MH	H	VH	H	VH	VH	MH	H	H	VH
C8	H	H	MH	MH	MH	H	H	H	MH	M
C9	VH	VH	VH	MH	VH	H	MH	H	VH	H

Table 11.4 Linguistic rating variable for evaluate the rating of alternatives w.r.t each criterion.

Criteria	Product	Decision makers evaluation									
		E1	**E2**	**E3**	**E4**	**E5**	**E6**	**E7**	**E8**	**E9**	**E10**
C1	**P1**	VG	G	G	G	VG	VG	G	VG	G	G
	P2	MG	G	VG	G	G	G	MG	G	MG	G
	P3	G	G	F	G	F	MG	MG	G	G	F
C2	**P1**	VG	VG	MG	VG	VG	G	MG	F	G	G
	P2	G	F	F	VG	G	MG	G	G	MG	VG
	P3	G	G	G	VG	MG	G	MG	MG	G	G
Criteria	**Product**	**Decision makers evaluation**									
		E1	**E2**	**E3**	**E4**	**E5**	**E6**	**E7**	**E8**	**E9**	**E10**
C3	**P1**	VG	G	MG	VG	VG	VG	MG	VG	G	G
	P2	VG	VG	G	MG	F	VG	VG	MG	VG	VG
	P3	VG	VG	G	VG	MG	G	VG	G	VG	VG
C4	**P1**	VG	VG	G	MG	VG	VG	VG	G	MG	G
	P2	MG	G	F	G	G	MG	VG	G	G	G
	P3	G	VG	MG	G	MG	G	MG	VG	MG	G

(*Continued*)

Table 11.4 Linguistic rating variable for evaluate the rating of alternatives w.r.t each criterion. (*Continued*)

C5	**P1**	G	MG	G	MG	VG	F	VG	G	G	VG
	P2	G	G	G	G	G	G	F	F	F	G
	P3	G	F	F	MG	G	F	G	F	VG	G
C6	**P1**	VG	G	MG	G	G	MG	G	G	G	MG
	P2	G	VG	G	MG	F	G	F	G	G	G
	P3	G	G	G	G	MG	G	G	G	MG	G
C7	**P1**	G	VG	G	G	VG	G	VG	MG	VG	G
	P2	MG	MG	VG	G	MG	MG	F	G	G	G
	P3	G	G	G	G	MG	MG	G	MG	G	G
C8	**P1**	G	VG	MG	VG	VG	G	G	VG	VG	VG
	P2	G	MG	G	G	F	G	MG	MG	G	MG
	P3	G	G	G	G	G	G	MG	G	VG	G
Criteria	**Product**	**Decision makers evaluation**									
		E1	**E2**	**E3**	**E4**	**E5**	**E6**	**E7**	**E8**	**E9**	**E10**
C9	P1	G	G	G	G	MG	MG	G	MG	G	G
	P2	G	G	G	G	G	G	F	F	F	G
	P3	G	G	G	MG	F	G	F	G	MG	G

Table 11.5 The fuzzy decision matrix and fuzzy weights of alternatives.

Criteria/alternative	C1	C2	C3
P1	(7.8,9.4,10)	(7.2,8.8,9.6)	(8.0,9.4,9.9)
P2	(6.4,8.3,9.5)	(6.0,8.0,9.3)	(7.4,8.8,9.5)
P3	(4.7,6.5,7.9)	(6.8,8.7,9.8)	(7.3,8.5,8.9)
Weight	(0.66,0.83,0.95)	(0.03,0.12,0.64)	(0.66,0.82,0.94)
Criteria/alternative	**C4**	**C5**	**C6**
P1	(7.6,9.1,9.8)	(7.4,9.1,9.9)	(6.6,8.5,9.7)
P2	(6.6,8.5,9.6)	(5.4,7.4,8.8)	(6.2,8.1,9.3)
P3	(5.05,6.53,7.65)	(4.85,6.33,7.35)	(5.25,6.83,7.85)
Weight	(0.8,0.93,0.98)	(0.34,0.54,0.74)	(0.7,0.86,0.96)
Criteria/alternative	**C7**	**C8**	**C9**
P1	(7.6,9.2,9.9)	(8,9.4,9.9)	(6.4,8.4,9.7)
P2	(5.8,7.7,9.2)	(6,7.9,9.3)	(5.8,7.8,9.1)
P3	(5.9,7.7,8.8)	(5.45,7.03,7.95)	(4.45,6.03,7.25)
Weight	(0.74,0.89,0.97)	(0.58,0.78,0.94)	(0.74,0.89,0.97)

The distance calculation of the alternatives w.r.t to fuzzy positive ideal solution and fuzzy negative ideal solution are shown in Table 11.8. The closeness coefficient calculations are shown in Table 11.9 to compute the best alternative as per given criteria, expert ratings and alternatives.

The calculations indicate that P3 is the best candidate among the alternatives.

Table 11.6 Fuzzy normalized decision matrix.

Criteria/alternative	C1	C2	C3
P1	(0.78,0.94,1)	(0.81,0.95,1)	(0.81,0.95,1)
P2	(0.64,0.83,0.95)	(0.75,0.89,0.96)	(0.75,0.89,0.96)
P3	(0.47,0.65,0.79)	(0.74,0.86,0.9)	(0.74,0.86,0.9)
Weight	(0.66,0.83,0.95)	(0.03,0.12,0.64)	(0.66,0.82,0.94)
Criteria/alternative	**C4**	**C5**	**C6**
P1	(0.78,0.93,1)	(0.75,0.92,1)	(0.68,0.88,1)
P2	(0.67.0.87,0.98)	(0.55,0.75,0.89)	(0.64,0.84,0.96)
P3	(0.52,0.67,0.78)	(0.49,0.64,0.74)	(0.54,0.7,0.81)
Weight	(0.8,0.93,0.98)	(0.34,0.54,0.74)	(0.7,0.86,0.96)
Criteria/alternative	**C5**	**C8**	**C9**
P1	(0.77,0.93,1)	(0.68,0.58,0.55)	(0.66,0.87,1)
P2	(0.59,0.78,0.93)	(0.91,0.69,0.59)	(0.60,0.80,0.94)
P3	(0.6,0.78,0.89)	(1,0.78,0.69)	(0.46,0.62,0.75)
Weight	(0.8,0.93,0.98)	(0.34,0.54,0.74)	(0.7,0.86,0.96)

Table 11.7 Fuzzy weighted normalized decision matrix.

Criteria/alternative	C1	C2	C3
P1	(0.51,0.78,0.95)	(0.02,0.11,0.64)	(0.53,0.78,94)
P2	(0.42,0.69,0.9)	(0.02,0.11,0.61)	(0.5,0.73,0.9)
P3	(0.31,0.54,0.75)	(0.02,0.1,0.58)	(0.49,0.71,0.85)
Criteria/alternative	**C4**	**C5**	**C6**
P1	(0.62,0.86,0.98)	(0.26,0.5,74)	(0.48,0.76,0.96)
P2	(0.54.0.81,0.96)	(0.19,0.41,0.66)	(0.45,0.72,0.92)
P3	(0.42,0.62,0.76)	(0.17,0.35,0.55)	(0.38,0.6,0.78)
Criteria/alternative	**C7**	**C8**	**C9**
P1	(0.57,0.83,0.97)	(0.39,0.45,0.52)	(0.49,0.77,0.97)
P2	(0.44,0.69,0.9)	(0.53,0.54,0.55)	(0.44,0.71,0.91)
P3	(0.44,0.69,0.86)	(0.58,0.61,0.65)	(0.34,0.55,0.73)

Table 11.8 Distance calculation.

Distances	A^*	A^-
P1	3.64	6.09
P2	3.97	5.71
P3	4.44	5.08

Table 11.9 Closeness coefficient.

CC_i	
P1	4.64
P2	4.97
P3	**5.44**

11.5 Conclusions

This chapter had enlisted the critical success factors that are to be investigated for institutions venturing into online teaching or looking for a change in their video conferencing software. The list id derived from first-hand experience of higher education institutions (HEI) that ventured into online teaching due to COVID-19 compulsions. This chapter further presents a mathematically simple Fuzzy TOPSIS method for selecting the video conferencing software. It would certainly help the implementers in selecting the best viable solution among the plethora of alternatives. The method also provides a mechanism for minimizing the cost criterion and maximizing the benefit criterion. Further, research directions in this area could be directed in terms of picking the most important CSF among the set of CSFs to be given due priority. Also, objectivity could be further enhanced by bringing the indecisiveness quotient also into the decision-making process. The same could be incorporated with the use of intuitionistic fuzzy sets.

References

1. Meinck, S, Fraillon, J., Strietholt, R., *The impact of the COVID-19 pandemic on education: International evidence from the Responses to Educational Disruption Survey, (REDS)*, UNESCO, 2022.
2. ILO Monitor 2nd Edition, in: *COVID-19 and the world of work: Updated estimates and analysis*, International Labor Organization, Geneva, 2020.
3. Hwang, C.L. and Yoon, K., *Multiple attributes decision making methods and applications*, Springer, Berlin, 1981.
4. Chen, C.T., Extensions of the TOPSIS for group decision-making under fuzzy environment. *Fuzzy Sets Syst.*, 114, 1–9, 2000.
6. Benitez, J.M., Martin, J., Roman, C., Using Fuzzy number for measuring quality of service in hotel industry. *Tour. Manage.*, 28, 544–555, 2007.
7. Wang, Y.M. and Elhag, T.M.S., Fuzzy TOPSIS method based on alpha level sets with an application to bridge risk assessment. *Expert Syst. Appl.*, 31, 309–319, 2005.
8. Wang, Y.J. and Lee, H.S., Generalizing TOPSIS for fuzzy multiple-criteria group decision-making. *Comput. Math. Appl.*, 53, 1762–1772, 2007.
9. Lin, C.T., Chen, C.B., Ting, Y.C., An ERP model for supplier selection in electronics industry. *Expert Syst. Appl.*, 38, 1760–1765, 2011.
10. Ertugrul, I. and Karakasoglu, N., Performance evaluation of Turkish cement firms with fuzzy analytic hierarchy process and TOPSIS methods. *Expert Syst. Appl.*, 36, 702–715, 2009.

12

Estimation of Nonperforming Assets of Indian Commercial Banks Using Fuzzy AHP and Goal Programming

Kandarp Vidyasagar* **and Rajiv Kr. Dwivedi**

University Department of Mathematics, Vinoba Bhave University, Hazaribagh, Jharkhand, India

Abstract

The present study computes the various parameters, which are different subdivis ions of the total nonperforming assets (NPA) of Indian Commercial Bank from the financial year 2013–2014 to 2019–2020. The variables were preferred on the several subdivisions. The secondary data have been collected from the audited annual reports of the State Bank of India. Multiple linear equations have been formulated, and Fuzzy Analytic Hierarchy Process (AHP) with extended Saaty's scale was used for weight estimation of the several parameters. Under these constraints, the goal programming model is developed, and the final result has been obtained using Lingo19. The study reveals that the various subdivisions of both the priority and nonpriority sectors of the Indian economy, which include agriculture and allied activities, industries (macro and small, medium and large), and personal loans have shown a moderate growth in order to increase the total NPA. On the contrary, the drastic growth has been observed in NPA of Services (priority and nonpriority sectors), which has crossed the model results. Further studies were suggested by partitioning the time period on a quarterly basis for the samples and calculating over the various microeconomic and macroeconomic factors and other bank-specific variables for more enhancement.

Keywords: Nonperforming assets, fuzzy AHP, goal programming

**Corresponding author*: vidyasagarkandarp@gmail.com

Sachi Nandan Mohanty, Prasenjit Chatterjee and Bui Thanh Hung (eds.) Fuzzy Computing in Data Science: Applications and Challenges, (215–236) © 2023 Scrivener Publishing LLC

12.1 Introduction

Analyzing a diversified set of objectives to obtain the best prediction is cumbersome. Achieving the target level is an out dare without decision making. Over the past few decades, various approaches and mathematical model have been developed to solve the real world problems. Goal programming (GP) is a robust method to solve such problems having multiple objectives. Initially, GP was introduced by Charnes and Cooper (1961) [1]. Later, it became very popular because of its simple method and easy application. GP is a branch of multiobjective optimization, which is a part of multicriteria decision analysis, which helps us in solving multiple conflicting objective measures. The basic aim of GP is to incorporate all goals in a single model that can further be solved simultaneously. The linear programming has a lousy efficacy over goal programming because of its computational efficiency [2].

In 1965, Zadeh propounded the fuzzy set theory [3]. Zimmerman used the membership function of fuzzy sets for the first time in goal programming [4, 5]. An ambient overview on the current study for fuzzy AHP and GP was discussed for optimization of bank liquidity [6]. Integrated method using fuzzy AHP and goal programming was developed for a supplier selection problem [7]. Fuzzy AHP has been used to asses a methodology for water management [8]. Fuzzy AHP and DEMATEL-ANP were combined to make a new hybrid multicriteria decision making model for selecting best space for free time in an urban site [9]. Weighted linear model for Fuzzy Goal Programming was developed, which could easily be used to solve real life problems [10]. Several researchers developed models for real time decision problem using fuzzy goal programming, such as vendor selection in supply chain, has been solved using the same approach [11]. Further, Fuzzy AHP and Fuzzy Goal Programming were together used for solving six primary goals in project selection, such as maximization of profit, customer satisfaction, processing capability along with minimization of cost, completion time, and risk [12].

Nonperforming assets (NPAs) are those credit facility in which the interest and instalment of principal remained due for a defined time period. NPA has been classified as substandard doubtful and loss category on the basis of various criteria speculated by Reverse Bank of India (RBI). A loan amount is considered as substandard assets if it remains nonperforming for a period less than or up to 1 year. Doubtful assets are those assets, which remained under substandard category for 1 year. Loss assets are those in which loss has been identified, but has not been fully written off. NPAs of Scheduled commercial banks (SCBs) increased from Rs 3, 23, 464 crore as on March 31, 2016 to Rs10, 6, 187 crore as on March 31, 2018. The main reason behind this drastic increase in NPAs was demonetization [13]. After 2018, there was a downfall in NPAs by Rs 1, 02, 562 crore and new figure in March 31, 2019, was Rs. 9, 33, 625 crore. The parlous state of NPA in Indian bank was already a matter of severe concern before COVID-19 struck.

The prolonged lockdown all over the country has certainly downturn the economic parameter and disrupted the demand and supply chain. The result was protracted slowdown of Indian economy and drastic rise in NPA of Indian Banks [14]. As per Reserve bank of India (RBI), in April 2020 several measures have been taken for liquidity management and the operating procedure of monetary policy [15].

Non-performing assets have significantly induced the fall in profit measures of Public Sector Banks in India [16]. NPA status in India was discussed over microdeterminants and macrodeterminants to report the rise in NPA [17]. The movement in NPA along with several macroeconomic indicators that would affect NPA were detailed with the primary objective of improving the quality of assets by reducing the NPA of Public Sector Banks [18].

Factor affecting nonperforming assets of commercial banks in India have been described. The factors were loan growth, cost effectiveness, bank size, etc. [19].

In the present paper, fuzzy AHP is applied for calculation of weight of various dependent variables of the nonperforming assets, which are agriculture and allied activities, industries (micro & small, medium & large), services & personal loans of both priority and nonpriority sectors. Further linear model is formed using these variables. Finally, goal programming model is developed for these set of constrains.

12.1.1 Basic Concepts of Fuzzy AHP and Goal Programming

The theory of fuzzy sets was defined by Zadeh in 1965. It defines a fuzzy set A by membership function $\mu_A(x)$ over the universe of discourse.

$$\mu_A : X \rightarrow [0,1]$$

So, the natural generalizations of general numbers are expressed as fuzzy numbers. Any ordinary numbers can be expressed as fuzzy number by using the following membership functions.

$$\mu_a(x)=\begin{cases} 1, & if x = a \\ 0, & if x \neq a \end{cases}$$

There are several types of fuzzy numbers like: triangular fuzzy number (TFN), trapezoidal fuzzy number (TrFn), and Gaussian fuzzy number (GFN). Out of these triangular fuzzy numbers are the simplest.

A fuzzy number on R is said to be triangular fuzzy number, if its membership function $\mu_A(x): R \rightarrow [0,1]$ is defined as:

$$\mu_A(x)=\begin{cases} \frac{x}{m-l}-\frac{l}{m-l}, & x\in[l,m] \\ \frac{x}{m-u}-\frac{u}{m-u}, & x\in[m,u] \end{cases}, \quad otherwise$$

The triangular fuzzy numbers are expressed as (l, m, u). Here, the parameter l represents the smallest possible value, m is the most possible value, and u is the largest possible value.

Out of various operations in triangular fuzzy numbers (l, m, u), two most important operations used in this study have been illustrated for (l_1, m_1, u_1) and (l_2, m_2, u_2)

The multiplication of TFN is defined as:

$$(l_1, m_1, u_1)(l_2, m_2, u_2) = (l_1 . l_2, m_1 . m_2, u_1 . u_2)$$

The inverse of TFN is defined as $(l_1, m_1, u_1)^{-1} = \left(\frac{1}{u_1}, \frac{1}{m_1}, \frac{1}{l_1}\right)$

Fuzzy AHP: Analytic hierarchy process (AHP) is one of the well-known tool used in multicriteria decision making. It was proposed by Saaty in 1980 [20]. The traditional tools, which were used for formal modelling, reasoning, and computing dichotomous deterministic data, were precise in character. But the complexities of the financial management system are always in a situation of risk and uncertainty prevailing in the business world. For this fuzzy, AHP was proposed by various authors.

Fuzzy AHP was introduced by Chang [21]. Let $X=\{x_1,x_2,\cdots,x_n\}$ be the object set and $G=\{g_1,g,\cdots,g_n\}$ be the goal set. Extent analysis for each objective is performed for all the goals, respectively. Let fuzzy extent analysis values calculated/determined are:

$$M^1_{gi}, M^2_{gi}, M^3_{gi}, \cdots M^n_{gi}\ ,\ i=1,2,3,\cdots n$$

where $M^j_{gi}\ (j=1,2,3\cdots,m)$ are triangular fuzzy numbers.

At first, the fuzzy synthetic extent is calculated by:

$$S_i=\sum_{j=1}^{m} M^j_{gi}\left[\sum_{i=1}^{n}\sum_{j=1}^{m} M^j_{gi}\right]^{-1} \tag{12.1}$$

$\sum_{j=1}^{m} M^j_{gi}$ is obtained by addition of m extent analysis values in the fuzzified pair wise comparison matrix.

$$\sum_{j=1}^{m} M^j_{gi}=\left(\sum_{j=1}^{m} l_j \sum_{j=1}^{m} m_j \sum_{j=1}^{m} u_j\right) \tag{12.2}$$

$\left[\sum_{i=1}^{n}\sum_{i=1}^{m} M^j_{gi}\right]^{-1}$ iscalculatedbyfuzzyadditionof $M^j_{gi}\,(j=1,2,3,\cdots,m)$ which is done as:

$$\sum_{i=1}^{n}\sum_{i=1}^{m} M^j_{gi}=\left(\sum_{i=1}^{n} l_i \sum_{i=1}^{n} m_i \sum_{i=1}^{n} u_i\right) \tag{12.3}$$

After calculation of the S_i value, the degree of possibility is calculated. If $M_1=(l_1,m_1,u_1)$ and $M_2=(l_2,m_2,u_2)$ are two triangular fuzzy numbers and if $M_2 \geq M_1$ then,

$$V(M_2 \geq M_1)=\begin{cases}1, & if\quad m_2 \geq m_1\\ 0, & if\quad l_1 \geq u_2\end{cases}\text{, Otherwise}\ \frac{l_1-u_2}{(m_2-u_1)-(m_2-l_2)} \tag{12.4}$$

Here, D is the highest point of intersection of μM_1 and μM_2. Figure 12.1 clearly states that to compare between two Triangular fuzzy numbers M_1 and M_2, we need both the values $V(M_1 \geq M_2)$ & $V(M_2 \geq M_1)$.

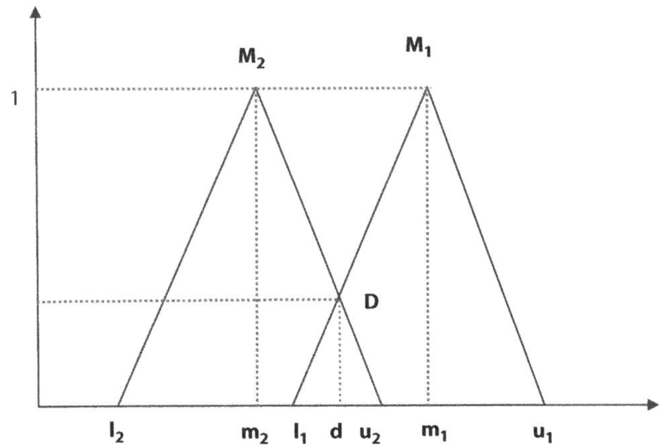

Figure 12.1 Intersection between two triangular fuzzy numbers (M_1 & M_2).

The degree of possibility of a convex fuzzy numbers to be greater than K convex fuzzy numbers $M_i \ (i=1,2,3,\cdots,k)$ is given by:

$$V(M \geq M_1, M_2, \cdots, M_k) = Min\ V\ (M \geq M_i),\quad i=1,2,\cdots,k$$

Let $d(A_i) = Min\ V\ (S_i \geq S_k)\ \ for\ k=1,2,3,\cdots,n\ ;k \neq i$
The weight vector is given by:

$$W' = (d'(A_1), d'(A_2), \cdots, d'(A_n))$$

where $A_i(\ i=1,2,3,\cdots,n)$ are n elements.

Lastly, the weight vectors are normalized to obtain the final weight;

$$W = (d\ (A_1), d(A_2), \cdots, d(A_n))$$

where W is a nonfuzzy number.

In earlier days, profit maximization was the only goal of financial management system. But now, there is no single universal goal. So, goal programming is an approach used for solving such multiple objective prioritized optimization problems.

The approach for formulating goal programming model is same as linear programming model. The decision variables are $x_1, x_2, \ldots, x_n$ first defined.

All specified goals are ranked in order of priority. A general goal programming can be stated as:

$$\text{Min } Z = \sum_{i=1}^{n} W_i \left(d_i^{+} + d_i^{-}\right) \;, \; i = 1,2,\cdots, m \tag{12.5}$$

$$\sum_{i=1}^{n} a_{ij} X_j \;+\; d_i^{-} - d_i^{+}, \;\; b_i = 1,2,\cdots, n$$

and

$$X_j,\; d_i^{-}, d_i^{+} \;\geq 0 \;,\; i = 1,2,\cdots, m \;; j = 1,2,\cdots, n$$

When W_i is weight vector calculated by Fuzzy AHP method. d_i^{+} & d_i^{-} are positive and negative deviations from the ith goal.

12.2 Research Model

The model has been developed to estimate the nonperforming assets on certain parameters of State Bank of India (SBI). The NPA of SBI has been broadly distributed among priority sector and nonpriority sector. Both these sectors have been individually classified into agriculture and allied activities, industries (micro and small, medium and large), services, and personal loans. Data of Table 12.1 has been taken from SBI annual report 2019–2020. Table 12.2 has been calculated weight of different criterion by applying fuzzy AHP with the help of Table 12.1. Table 12.3 has prepared by using original Saaty's scale which citation is provided below [20]. Table 12.4 has been prepared by me using original Saaty's scale mentioned in Table 12.3. Table 12.5 has been calculate with the help of Table 12.2 by applying Extended Saaty's scale, as in Table 12.4. Table 12.6 has been prepared using data from audited annual report of SBI from year 2013 to 2020. Then, the dependency of various sectors on gross NPA is calculated by applying fuzzy AHP to find the weights of the different criterion.

Now, the priority comparison matrix is made by simply dividing the percentage value of different criterion. For example, for the cell a_{12}, we have to find how dominant is agricultural NPA over the NPA in industries of priority sectors. Then, we will simply divide the percentage value of gross NPA to total NPA in agricultural sector by the percentage value of gross NPA to total NPA in the industrial sector.

After making the pairwise comparison matrix, the crisp value in the matrix are converted into fuzzy numbers using extension of Saaty's scale.

Table 12.1 Gross NPA in priority and non-priority sectors.

Sectors	Outstanding advances (Rs in Crore)	Gross NPA (Rs in Crore)	Percentage of gross NPA to total advances to that sectors
A. Priority sectors			
(i) Agriculture and allied activities	2,04,185.71	32,558.27	15.95
(ii) Industries (micro and small, medium and large)	1,01,080.54	18,738.88	18.54
(iii) Services	83,870.61	5,289.20	6.31
(iv) Personal loans	1,66,800.34	3,131.18	1.88
Subtotal	5,55,937.20	59,717.53	10.74
B. Non-priority sectors			
(i) Agriculture and allied activities	2,235.29	229.81	10.28
(ii) Industries (micro and small, medium and large)	10,54,285.42	74,644.63	7.08
(iii) Services	2,21,642.21	9,686.06	4.37
(iv) Personal loans	5,88,744.65	4,813.82	0.82
Subtotal	18,66,907.57	89,314.32	4.79
Total (A+B)	24,22,844.77	1,49,091.85	6.15

Table 12.2 Interdependency of different economic sectors.

	Agriculture and allied activities	Industries (micro and small, medium and large)	Services	Personal	Agriculture and allied activities	Industries (microand small, medium and large)	Services	Personal
Agriculture and allied activities	1	0.860	2.527	8.484	1.551	2.252	3.649	19.451
Industries (micro and small, medium and large)	1.162	1	2.938	9.861	1.803	2.618	4.242	22.609
Services	0.395	0.340	1	3.356	0.613	0.891	1.443	7.695
Personal	0.117	0.101	0.297	1	0.182	0.265	0.430	2.292
Agriculture and allied activities	0.644	0.554	1.629	5.468	1	1.451	2.352	12.536
Industries (micro and small, medium and large)	0.443	0.381	1.122	3.765	0.688	1	1.620	8.634
Services	0.273	0.235	0.692	2.324	0.425	0.617	1	5.329
Personal	0.051	0.044	0.129	0.436	0.079	0.115	0.181	1

Table 12.3 Original Saaty's scale for pairwise comparison.

Crisp value	Definition	Fuzzified value
1	Equal importance	(1, 1, 1+δ)
3	Weak dominance	(3- δ, 3, 3+ δ)
5	Strong dominance	(5- δ,5,5+ δ)
7	Demonstrated dominance	(7- δ, 7, 7+ δ)
9	Absolute dominance	(9- δ, 9, 9+ δ)
2,4,6,8	Intermediate values	(x-1,x,x+1) ; where x=2,4,6,8

δ is fuzzy distance(0.5≥ δ≤ 2).

Table 12.4 Extension of Saaty's scale.

Crisp value	Definition	Fuzzified value
1	Equal importance	(1,1,1+δ)
3	Weak dominance	(3- δ,3,3+ δ)
5	-	(5- δ,5,5+ δ)
7	-	(7- δ,7,7+ δ)
9	-	(9- δ,9,9+ δ)
11	-	(11- δ,11,11+ δ)
13	-	(13- δ,13,13+ δ)
15	-	(15- δ,15,15+ δ)
17	-	(17- δ,17,17+ δ)
19	-	(19- δ,19,19+ δ)
21	-	(21- δ,21,21+ δ)
23	-	(23- δ,23,23+ δ)
25	Absolute Dominance	(25- δ,25,25+ δ)

Table 12.5 Fuzzified matrix.

(1, 1, 1.5)	**(0.602, 0.861, 1.511)**	**(2.027, 2.527, 3.027)**	**(7.984,8.484, 8.984)**	**(1.051, 1.551, 2.051)**	**(1.752, 2.252, 2.752)**	**(3.149, 3.694, 4.149)**	**(18.951, 19.451, 19.951)**
(0.002,1.162,1.662)	(1, 1, 1.5)	(2.438, 2.938, 3.438)	(9.361, 9.861, 10.361)	(1.303, 1.803, 2.303)	(2.118, 2.618, 3.118)	(3.742, 4.242, 4.742)	(22.109, 22.609, 23.109)
(0.330,0.396,0.493)	(0291, 0.340, 0.410)	(1, 1, 1.5)	(2.856, 3.356, 3.856)	(0.470, 0.614, 0.886)	(0.617, 0.891, 1.608)	(0.943, 1.443, 1.943)	(7.195, 7.695, 8.195)
(0.111, 0.118,0.125)	(0.097,0.101, 0.107)	(0.259,0.298, 0.350)	(1, 1, 1.5)	(0.168, 0.183, 0.201)	(0.234, 0.266, 0.306)	(0.354, 0.430, 0.548)	(1.792, 2.292, 2.792)
(0.488,0.645,0.951)	(0.434, 0.555, 0.767)	(1.129, 1.629, 2.129)	(4.968, 5.468, 5.968)	(1, 1, 1.5)	(0.951, 1.451, 1.951)	(1.852,2.352,2.852)	(12.036,12.536,13.036)
(0.363,0.444,0.571)	(0.321, 0.382, 0.472)	(0.622, 1.122, 1.622)	(3.265, 3.765, 4.265)	(0.513,0.689,1.052)	(1,1,1.5)	(1.120,1.620,2.120)	(8.134,8.634,9.134)
(0.241,0.274,0.318)	(0.211, 0.236, 0.267)	(0.515, 0.693, 1.060)	(1.824, 2.324, 2.824)	(0.351,0.425,0.540)	(0.472,0.617,0.893)	(1,1,1.5)	(4.829,5.329,5.829)
(0.050,0.051,0.053)	(0.043, 0.044, 0.045)	(0.122, 0.130, 0.139)	(0.358, 0.436, 0.558)	(0.077,0.080,0.083)	(0.109,0.116,0.123)	(0.172,0.188,0.207)	(1,1,1.5)

$$x \in (1,25) - (1,3,5,7,9,11,13,15,17,19,21,23,25)(x - \delta, x, x + \delta)$$

$$\delta = (0.5 \le \delta \le 1)$$

With the help of this scale a new fuzzified matrix is created, are as follows in Table 12.5:

Using this fuzzified matrix, fuzzy synthetic extent is calculated.
is calculated.

$$A_1 = (36.516, 39.775, 43.952)$$
$$A_2 = (42.733, 46.233, 50.233)$$
$$A_3 = (13.702, 15.735, 18.891)$$
$$A_4 = (4.015, 4.688, 5.929)$$
$$A_5 = (22.858, 22.636, 29.154)$$
$$A_6 = (15.338, 17.656, 20.736)$$
$$A_7 = (9.433, 10.898, 13.231)$$
$$A_8 = (1.931, 2.045, 2.708)$$

$$\sum_{i=1}^{n} \sum_{j=1}^{m} M_{g_i}^{j} = (146.536, 162.666, 184.807)$$

$$\left[\sum_{i=1}^{n} \sum_{j=1}^{m} M_{g_i}^{j}\right]^{-1} = (0.0054, 0.0061, 0.0068)$$

Finally, using equation (12.1) Fuzzy synthetic extent is obtained as:

$$S_i = \sum_{j=1}^{m} M_{g_i}^{j} \left[\sum_{i=1}^{n} \sum_{j=1}^{m} M_{g_i}^{j}\right]^{-1}$$

$$S_1 = (0.197, 0.243, 0.299)$$
$$S_2 = (0.231, 0.282, 0.342)$$
$$S_3 = (0.073, 0.046, 0.128)$$

$$S_4 = (0.022, 0.29, 0.040)$$
$$S_5 = (0.123, 0.156, 0.198)$$
$$S_6 = (0.083, 0.108, 0.141)$$
$$S_7 = (0.051, 0.066, 0.090)$$
$$S_8 = (0.010, 0.012, 0.018)$$

Here weights have calculated by de-fuzzifying the Si values, i.e., the fuzzy synthetic extent values.

So weights are:

$$W' = [\ 0.246, 0.285, 0.099, 0.30, 0.159, 0.111, 0.069, 0.013]^T$$

Hence, normalized weights are:

$$W = [0.243, 0.284, 0.097, 0.028, 0.157,\ 0.110,\ 0.068, 0.012]$$

12.2.1 Average Growth Rate Calculation

The average growth rate in different sectors has been calculated using the data of seven years from 2013–2014 to 2019–2020. At first, the percentage increase between two consecutive years is calculated and then the average percentage increase in gross NPA is calculated by taking the mean of the percentage increase of different consecutive years. Table 12.6 shows the average growth rate of different sectors in percentage.

The first group of models input and output variable are taken in Table 12.7.

In Table 12.8, priority arrangement of variables is done as per the normalized weights calculated by Fuzzy AHP.

The linear problem on the basis of obtained weights is discussed below in equation 12.6. The lower limit in different equations has been taken from Table 12.1 from the column of gross NPA. The upper limit in the equation has been obtained by calculating the average percentage increase over gross NPA in 2020. For example, for x_1, i.e., NPA from industry (priority sector) the gross NPA is taken always greater than that in 2020, which is Rs. 18738.88 cr.

$$x1 > 18738.88$$

Table 12.6 Average growth rate in different sectors.

Gross NPA in subsequent years								Gross NPA in %age						
Sectors	2013–2014	2014–2015	2015–2016	2016–2017	2017–2018	2018–2019	2019–2020	% increase from 2013–2014 to 2014–2015	% increase from 2014–2015 to 2015–2016	% increase from 2015–2016 to 2016–2017	% increase from 2016–2017 to 2017–2018	% increase from 2017–2018 to 2018–2019	% increase from 2018–2019 to 2019–2020	AVG % increase in Gross NPA
Priority sectors														
NPA from Agriculture & allied activities	10327.11	10216.74	9839.11	7354.6	20964.77	23335.8	32558.27	−1.069	−3.696	−25.251	185.055	11.310	39.521	34.312
NPA from Industry (Micro & small, medium & Large)	4174.09	7087.13	11602.3	11536	16020.84	12545.6	18738.88	69.789	63.709	−0.571	38.877	−21.692	49.366	33.246
NPA from Services	2424.62	1699.94	1747.36	2378.6	7339.66	9674.48	5289.2	−29.888	2.790	36.122	208.577	31.811	−45.328	34.014
Personal Loan	1659.29	1202.51	1033.79	972.64	3332.33	2882.01	3131.18	−27.529	−14.031	−5.915	242.607	−13.514	8.646	31.711
Non-priority sectors														
NPA agriculture & allied activities	224.86	199.91	496.94	99.26	301.93	89	229.81	−11.096	148.582	−80.026	204.181	−70.523	158.213	58.222
NPA from Industry (Micro & small, medium & Large)	34566.58	31152.77	67674.8	82086	162785	112412	74644.63	−9.876	117.235	21.295	98.309	−30.945	−33.597	27.070
NPA from Services	6977.35	4014.14	4355.62	6704.7	9264.85	8007.3	9686.06	−42.469	8.507	53.933	38.184	−13.573	20.965	10.924
Personal Loan	1679.34	1152.2	1422.93	1210.8	3418.09	3804.5	4813.82	−31.390	23.497	−14.911	182.312	11.305	26.530	32.890
Total NPA	62033.24	56725.34	98172.8	112343	223427.5	172750	149091.9	−8.557	73.067	14.434	98.880	−22.682	−13.695	23.575

Table 12.7 Arrangement of priority and non-priority sectors as per weight.

Priority sector		Input & output variable
I	NPA from Agriculture & allied activities	X_2
II	NPA from Industry (Micro & small, medium & Large)	X_1
III	NPA from Services	X_5
IV	Personal Loan	X_7
Non-priority sector		
V	NPA agriculture & allied activities	X_3
VI	NPA from Industry (Micro & small, medium & Large)	X_4
VII	NPA from Services	X_6
VIII	Personal Loan	X_8

Table 12.8 Priority arrangement.

Sectors		Priority weight after normalization	Coefficient
I	NPA from Industry (Micro & small, medium & Large) (priority sector)	X_1	0.284
II	NPA from Agriculture & allied activities (priority sector)	X_2	0.243
III	NPA agriculture & allied activities (Non-priority sector)	X_3	0.157
IV	NPA from Industry (Micro & small, medium & Large) (Non-priority sector)	X_4	0.110
V	NPA from Services (priority sector)	X_5	0.097
VI	NPA from Services (Non-priority sector)	X_6	0.068
VII	Personal Loan (priority sector)	X_7	0.028
VIII	Personal Loan (Non-priority sector)	X_8	0.012

Now, for the upper limit of x_1, average growth rate is 33.246%. Hence,

$$x_1 < 18738.88 + 33.246\% \text{ of } 18738.88$$

$$x_1 < 18738.88(1 + 0.33246)$$

$$x_1 < 18738.88 * 1.33246$$

Similarly, the upper and lower limits for x_2, x_3, x_4, x_5, x_6, x_7, and x_8 can be calculated to get equation 12.6.

$$\begin{gathered}
18738.88<x_1< 1.33246*18738.88 \\
32558.27<x_2<1.34312*32558.27 \\
229.81<x_3<1.58522*229.81 \\
74644.63<x_4<1.27070*74644.63 \\
5289.20>x_5>1.34014*5289.20 \\
9686.06<x_6<1.10924*9686.06 \\
3131.18>x_7>1.31711*3131.18 \\
4813.82>x_8>1.32890*4813.82 \\
149091.85 < \sum_{j=1}^{8} X_j < 1.23575 * 149091.85
\end{gathered} \tag{12.6}$$

Equivalent goal programming model is as follows:

Here, the coefficients in z have been taken from Table 12.8, which is the normalized weight.

z=0.284*(d1plus)+0.243*(d2plus)+0.157*(d3plus)+0.110*(d4plus)+0.097*(d5plus)+0.068*(d6plus)+0.028*(d7plus)+0.012*(d8plus)

$$x1>18738.88$$

$$x1-1.33246*18738.88+d1plus=0$$

$$x2>32558.27$$

$$x2-1.34312*32558.27+d2plus=0$$

$$x3>229.81$$

$$x3-1.58522*229.81+d3plus=0$$

$$x4>74644.63$$

$$x4-1.27070*74644.63+d4plus=0 \tag{12.7}$$

$$x5>5289.20$$

$$x5-1.34014*5289.20+d5plus=0$$

$$x6>9686.06$$

$$x6-1.10924*9686.06+d6plus=$$

$$x7>3131.18$$

$$x7-1.31711*3131.18+d7plus=0$$

$$x8>4813.82$$

$$x8-1.32890*4813.82+d8plus=0$$

$$x1+x2+x3+x4+x5+x6+x7+x8<1.23575*149091.85$$

$$x1+x2+x3+x4+x5+x6+x7+x8>149091.85$$

Outcome recorded after minimizing the Goal programming problem (12.7) using Lingo19 optimization software clearly shows that the global optimal solution is possible as in Figure 12.2.

Variable	Value	Reduced Cost
D1PLUS	0.000000	0.1740000
D2PLUS	0.000000	0.1330000
D3PLUS	0.000000	0.4700000E-01
D4PLUS	2593.709	0.000000
D5PLUS	1799.068	0.000000
D6PLUS	1058.105	0.000000
D7PLUS	992.9285	0.000000
D8PLUS	1583.265	0.000000
X1	24968.81	0.000000
X2	43729.66	0.000000
X3	364.2994	0.000000
X4	92257.22	0.000000
X5	5289.200	0.000000
X6	9686.060	0.000000
X7	3131.180	0.000000
X8	4813.820	0.000000

Row	Slack or Surplus	Dual Price
1	578.5699	-1.000000
2	6229.928	0.000000
3	0.000000	-0.1100000
4	11171.39	0.000000
5	0.000000	-0.1100000
6	134.4894	0.000000
7	0.000000	-0.1100000
8	17612.59	0.000000
9	0.000000	-0.1100000
10	0.000000	-0.1300000E-01
11	0.000000	-0.9700000E-01
12	0.000000	-0.4200000E-01
13	0.000000	-0.6800000E-01
14	0.000000	-0.8200000E-01
15	0.000000	-0.2800000E-01
16	0.000000	-0.9800000E-01
17	0.000000	-0.1200000E-01
18	0.000000	0.1100000
19	35148.40	0.000000

Figure 12.2 Feasibility result of goal programming model.

12.3 Result and Discussion

The proposed model has estimated the NPA of priority and nonpriority sectors which includes agriculture and allied activities, industries (micro and small, medium and large), services and personal loans.

A comparison table result has been mentioned below in Table 12.9, which clearly shows that the NPA in agriculture and allied activities, industries (micro and small, medium and large) and personal loans of both priority and nonpriority sector has shown modest growth but are below the benchmark of the average growth rate as per the model result. However,

Table 12.9 Comparison of gross NPA in SBI annual report 2020–2021 to gross NPA as per model result.

Sectors	**Gross NPA in SBI annual report 2020-21 (Rs in Crore)**	**Gross NPA as per model result (Rs in Crore)**
Priority sector		
(i) Agriculture & allied activities	32,392.47	43,729.66
(ii) Industries (micro & small, medium & large)	11,206.95	24,968.81)
(iii) Services	**10,198.53**	**5,289.200**
(iv) Personal loans	2,352.84	3,131.180
Non-priority sector		
(i) Agriculture & allied activities	205.85	364.2994
(ii) Industries (micro & small, medium & large)	47,770.41	92,257.22
(iii) Services	**17,636.56**	**9,686.060**
(iv) Personal loans	4,625.41	4813.820
Total	1,26,389.02	184,240.25

the NPA in services has almost doubled showing the rapid growth compared to the average growth of the model result.

12.4 Conclusion

In view of the outcome of the proposed model, all eight different goals have been scrutinized and achieved. The findings clearly state that the performance of state bank of India (SBI) is good in agriculture and allied activities, industries (micro and small, medium and large) and personal loans of both priority and nonpriority sector but the NPA of the service sector has doubled with respect to the model result. This shows a serious concern of the bank to work individually on service sector in order to maintain its NPA. The infeasibility for the linear model was overcome by the goal programming model. Furthermore, the developed model can be implemented to prepare a solution system, which can help banks and other financial institutions to design a blueprint. This can further help them to identify their benchmark, which can be achieved in the future.

References

1. Charnes, A. and Cooper, W.W., Management models and industrial applications of linear programming. *Manage. Sci.*, *4*, 1, 38–91, 1957.
2. Aouni, B. and Kettani, O., Goal programming model: A glorious history and a promising future. *Eur. J. Oper. Res.*, *133*, 2, 225–231, 2001.
3. Zadeh, L.A., Fuzzy sets, in: *Fuzzy Sets, Fuzzy Logic, and Fuzzy Systems: Selected Papers*, Zadeh, L.A. (Ed.), pp. 394–432, 1996.
4. Zimmermann, H.J., Description and optimization of fuzzy systems. *Int. J. Gen. Syst.*, *2*, 1, 209–215, 1975.
5. Zimmermann, H.J., Fuzzy programming and linear programming with several objective functions. *Fuzzy Sets Syst.*, *1*, 1, 45–55, 1978.
6. Mohammadi, R. and Sherafati, M., Optimization of bank liquidity management using goal programming and fuzzy AHP. *Res. J. Recent Sci.*, 4, 6, 53–61, 2015.
7. Sivrikaya, B.T., Kaya, A., Dursun, M., Çebi, F., Fuzzy AHP–goal programming approach for a supplier selection problem. *Res. Logist. Prod.*, *5*, 271–285, 2015.
8. Srdjevic, B. and Medeiros, Y.D.P., Fuzzy AHP assessment of water management plans. *Water Res. Manage.*, *22*, 7, 877–894, 2008.
9. Pourahmad, A., Hosseini, A., Banaitis, A., Nasiri, H., Banaitienė, N., Tzeng, G.H., Combination of fuzzy-AHP and DEMATEL-ANP with GIS in a new

hybrid MCDM model used for the selection of the best space for leisure in a blighted urban site. *Technol. Econ. Dev. Econ.*, *21*, 5, 773–796, 2015.
10. Yaghoobi, M.A., Jones, D.F., Tamiz, M., Weighted additive models for solving fuzzy goal programming problems. *Asia Pac. J. Oper. Res.*, 25, 05, 715–733, 2008.
11. Kumar, M., Vrat, P., Shankar, R., A fuzzy goal programming approach for vendor selection problem in a supply chain. *Comput. Ind. Eng.*, *46*, 1, 69–85, 2004.
12. Kahraman, C. and Büyüközkan, G., A Combined fuzzy AHP and fuzzy goal programming approach for effective six-sigma project selection. *J. Mult.-Valued Log. Soft Comput.*, *14*, 6, 599–615, 2008.
13. Vidyasagar, K. and Dwivedi, R.K., Post demonetisation effect on banking sector, micro financing institutions (MFI) and job opportunities. *Int. J. Manage. Innov. Entrep. Res.*, 6, 2, 39–44, 2020, https://doi.org/10.18510/ijmier.2020.624.
14. Vekariya, S.P., Can Indian banking industry overcome NPA issue uring the pandemic time? *PBME*, 103, 103–110.
15. *Monetary policy report*, Reserve Bank of India (RBI), India, April 2020.
16. Dave, P. and Joshi, Y.C., *NPA and indian banking sector: Assessing the impact on the profitability of public sector banks in India*, GH Patel Postgraduate Institute of Business Management, Gujrat, India, 1935.
17. Khandelwal, V. and Modi, S., *India's NPA situation with an emphasis on public and private banks*, vol. 3857444, SSRN, Haryana, India, 2021.
18. Haralayya, D., Study on non performing assets of public sector banks. *Iconic Res. Eng. J. (IRE)*, *4*, 12, 52–61, 2021.
19. Prasanth, S., Nivetha, P., Ramapriya, M., Sudhamathi, S., Factors affecting non performing loan in India. *Int. J. Sci. Technol. Res.*, *9*, 1, 1654–1657, 2020.
20. Saaty, T., *The analytic hierarchy process (AHP) for decision making*, pp. 1–69, Kobe, Japan, 1980.
21. Chang, D.Y., Applications of the extent analysis method on fuzzy AHP. *Eur. J. Oper. Res.*, *95*, 3, 649–655, 1996.

13

Evaluation of Ergonomic Design for the Visual Display Terminal Operator at Static Work Under FMCDM Environment

Bipradas Bairagi

Department of Mechanical Engineering, Haldia Institute of Technology, Haldia, India

Abstract

This study explores a new fuzzy-based mathematical algorithm in analysis and evaluation of performance characteristics of static work station for accurate and right decision making. In this technique, diverse important criteria are identified and their importance weights are estimated by applying experience and opinion of the experts involved in the decision making process. Based on the criteria, sufficient numbers of alternative work stations are screened, and respective performance ratings are evaluated for further processing. The experts also assign relative importance to the selection criteria on the basis of their experience, knowledge, and perception. Thereafter, performance ratings of alternative work stations, weight of criteria, and importance weight of users are combined together with logical and soft-computing tool proposed in the study. The investigation found that the proposed method is useful for evaluation and selection of ergonomic design for static work under fuzzy multicriteria decision making environment.

***Keywords*:** Ergonomic workplace evaluation, decision making, FMCDM, soft computing techniques

Email: bipradasbairagi79@gmail.com

Sachi Nandan Mohanty, Prasenjit Chatterjee and Bui Thanh Hung (eds.) Fuzzy Computing in Data Science: Applications and Challenges, (237–252) © 2023 Scrivener Publishing LLC

13.1 Introduction

The global market is constantly changing. The taste and preference of customers also vary at a rapid pace. Capturing the trend of customers' or end users' choice is now a crucial task for the managers of manufacturing organizations. Selection of suitable ergonomic design for the visual display terminal operators at static work under fuzzy multicriteria decision making environment is critical for the concerned decision makers. The selection of appropriate ergonomic design involves several alternatives, under multiple criteria considerations. Hence, such decision making procedure deserves implementation of suitable MCDM approach. If the decision is to make under subjective criteria associated with vagueness, imprecision, and ambiguity, then application of fuzzy set theory in combination with MCDM techniques is inevitable. A literature survey has been conducted on application of MCDM for decision making in ergonomic design selection that explores the literature gap.

Jain *et al.* applied an integrated multiple criteria decision making method for determining the risk level of musculoskeletal disorders among users of handled apparatus [1]. Eraslan *et al.* implemented a multiple-criteria decision making approach for the evaluation and selection of suitable design of office layout [2]. Mohanty *et al.* proposed a new fuzzy-based multiple attribute decision making method for the purpose of identifying the best product that conform ergonomic related considerations. The method was illustrated with the solution of a problem on office chair selection [3]. Halder *et al.* employed an integrated multiple criteria decision making approach for evaluation and selection of seat comfort with ergonomic considerations [4]. Mohammad *et al.* combined multiple criteria decision making technique with ergonomic technique for the purpose of risk measurement. The approach has been applied to assess the repetitive task [5].

A participatory approach is often the most effective way to decide the product features and accessories that are likely to have the greatest cost benefits for different groups of workers in an organization. Document holders, racks are often of great benefits for different groups of workers. There are many tools available for assessment of workstations where static work is carried out. Rapid Entire Body Assessment (REBA) is a postural analysis tool that gives a composite risk rating [6].

Postural analysis is a prevailing method for evaluating work actions. Musculoskeletal risk injury related to the postures (recorded), in the perspective of a complete ergonomic workplace appraisal, is a main aspect for executing change. A good number of postural analysis procedures include two, generally conflicting, qualities of generalization and compassion [7]. High generalization in a postural analysis technique may be compensated by little compassion. A requirement was professed within the range of postural analysis tools, particularly with sensitivity to the sort of volatile working postures found in health care and supplementary service industries. It directs to the growth of the postural analysis tool named Rapid Entire Body Assessment, REBA [8, 9].

Existing student's workstation within classroom facilities does not fully support an up-to-date need of contemporary academic regulations. Several researchers have initiated appropriate student chair design following the standard physiological dimensions [10–12]. However, their contributions are mostly related toward the theoretical level and mainly in two-dimensional and not three-dimensional geometric space. Basically, existing users' chairs are designed mostly in a traditional way and are not much considered according to ergonomics principles [13]. One of the most important needs for workspace chair design is to consider the relationship between the users' physical standards and chair's functional dimensions [14–16].

Production and inventory costs are important consideration in designing visual display workspace, but at the same time, it is also important to relate anthropometric data and ergonomics during design [17–19]. Extensive researches in the area of school chair and desk have been conducted in the Scandinavian countries including Denmark and Sweden and the drift is also noticed in Germany, France and Switzerland [20].

The initial point to design and develop an ergonomic users' chair, visual display space is to explore how the dimensions and styles of the chairs and desks are essentially used in the working environment to replicate the body dimensions and the functional requirements of present days' huge working population [21, 22]. Incorporating multicriteria-based soft computing technique for safe product design considering ergonomic point of views is newly introduced in this study. Basically ergonomic design is to consider the human factor and man-machine interface.

The analysis of the relevant literature survey reveals that fuzzy multicriteria decision making approach have not been used so far to evaluate and select ergonomic alternative design, at least at a satisfactory level. Multicriteria decision making process using fuzzy set theory may be a potential tool in this purpose. The manufacturing organizations are facing the problem of proper decision making regarding the selection of appropriate alternatives of ergonomic designs. It is evident from the above literature survey that suitable MCDM approach is necessary for making appropriate decision involving ergonomic considerations.

The objective of the current investigation is to develop a decision making framework that can be an essential tool in solving problem regarding selection of ergonomic design for the visual display terminal operator at static work under FMCDM environment.

The rest of the chapter is divided into some sections as follows. Section 13.2 proposes a new algorithm. Section 13.3 describes an illustrative example on ergonomic design evaluation and significant discussions. Lastly section 13.4 highlights some essential conclusions and direction of future research.

13.2 Proposed Algorithm

This research work proposes a novel algorithm consisting of 10 interrelated steps under FMCDM environment as described below.

Step 1: The algorithm starts with the formation of decision matrix comprising of performance rating for the alternatives with respect to decision making criteria. The performance ratings of alternatives are assessed by users, experts, or decision makers in linguistic variables. The decision matrix is constructed as follows.

$$M_d(P_j) = \begin{array}{c} \\ A_1 \\ \cdots \\ A_i \\ \cdots \\ A_m \end{array} \begin{array}{c} \begin{array}{ccccc} C_1 & \cdots & C_j & \cdots & C_n \end{array} \\ \begin{bmatrix} LV_{11} & \cdots & LV_{1j} & \cdots & LV_{1n} \\ \cdots & \cdots & \cdots & \cdots & \cdots \\ LV_{i1} & \cdots & LV_{ij} & \cdots & LV_{in} \\ \cdots & \cdots & \cdots & \cdots & \cdots \\ LV_{m1} & \cdots & LV_{mj} & \cdots & LV_{mn} \end{bmatrix} \end{array} \quad (13.1)$$

LV_{ij} denotes linguistic variable expressing performance rating of alternative A_i with respect to criterion C_j estimated by user or experts E_j. n is number of criteria and m is number of alternatives. Linguistic variables have five degrees which are transformed into five corresponding triangular fuzzy numbers (TFNs) as represented in Table 13.1.

Step 2: Formation of weight matrix for assigning importance weight to the criteria estimated by a committee of experts, decision makers and users. The proposed weight matrix for measuring criteria importance is formed as follows.

Table 13.1 Five degrees of linguistic variables for assessing subjective attributes.

Description	Abbreviation	Triangular fuzzy number
Very comfortable	VC	(80,90,100)
Comfortable	C	(60,70,80)
Average	A	(40,50,60)
Not comfortable	NC	(20,30,40)
Very uncomfortable	VU	(0,10,20)

$$M_d(P_j) = \begin{array}{c} \\ P_1 \\ \cdots \\ P_i \\ \cdots \\ P_r \end{array} \begin{array}{c} \begin{array}{ccccc} C_1 & \cdots & C_j & \cdots & C_n \end{array} \\ \begin{bmatrix} Lvw_{11} & \cdots & Lvw_{1j} & \cdots & Lvw_{1n} \\ \cdots & \cdots & \cdots & \cdots & \cdots \\ Lvw_{i1} & \cdots & Lvw_{ij} & \cdots & Lvw_{in} \\ \cdots & \cdots & \cdots & \cdots & \cdots \\ Lvw_{r1} & \cdots & Lvw_{rj} & \cdots & Lvw_{rn} \end{bmatrix} \end{array} \tag{13.2}$$

$M_{d(P_i)}$ is the decision matrix formed by the experts P_i. Lvw_{ij} denotes the linguistic variables for weights of criterion C_j, assessed by ith experts or user, where n is number of criteria and r is number of experts, decision makers, or users. Description of 5 degrees of linguistic variables, abbreviation, and corresponding TFN for weights are presented and described in the following Table 13.2.

Table 13.2 Linguistic variables, abbreviation and TFN for weights.

Description	Abbreviation	Triangular fuzzy number
Extremely important	EI	(0.8,0.9,1)
Very important	VI	(0.6,0.7,0.8)
important	I	(0.4,0.5,0.6)
Less-important	LI	(0.2,0.3,0.4)
Unimportant	UI	(0, 0.1,0.2)

Step 3: Conversion of linguistic terms of decision matrix into triangular fuzzy numbers (TFN) as per proposed conversion table as presented in step 2. A decision matrix comprising TFNs is represented as follows.

$$
M_{FD(P_i)} = \begin{array}{c} \\ A_1 \\ \cdots \\ A_i \\ \cdots \\ A_r \end{array}
\begin{array}{c} \begin{array}{ccccc} C_1 & \cdots & C_j & \cdots & C_n \end{array} \\
\begin{bmatrix}
(l_{11}, m_{11}, u_{11}) & \cdots & (l_{1j}, m_{1j}, u_{1j}) & \cdots & (l_{1n}, m_{1n}, u_{1n}) \\
\cdots & \cdots & \cdots & \cdots & \cdots \\
(l_{i1}, m_{i1}, u_{i1}) & \cdots & (l_{ij}, m_{ij}, u_{ij}) & \cdots & (l_{in}, m_{in}, u_{in}) \\
\cdots & \cdots & \cdots & \cdots & \cdots \\
(l_{m1}, m_{m1}, u_{m1}) & \cdots & (l_{mj}, m_{mj}, u_{mj}) & \cdots & (l_{mn}, m_{mn}, u_{mn})
\end{bmatrix} \end{array}
\tag{13.3}
$$

$M_{FD(P_i)}$ is the fuzzy decision matrix form by experts P_i. Here, (l_{ij}, m_{ij}, u_{ij}) denotes performance rating in triangular fuzzy number for alternatives A_i with respect to criteria C_j and assessed by ith user (expert) in the specified five degrees of gradation. The number of decision matrix comprising TFN is equal to the number of decision matrix comprising linguistic variables. In other words, the number of decision matrices comprising TFN is equal to the users or evaluators.

Step 4: Conversion of linguistic terms of the weight matrix into triangular fuzzy numbers (TFN) is accomplished as per proposed conversion table as presented in step 2. The weight matrix comprising TFNs is represented as follows.

$$
M_{FW} = \begin{array}{c} \\ P_1 \\ \cdots \\ P_i \\ \cdots \\ P_r \end{array}
\begin{array}{c} \begin{array}{ccccc} C_1 & \cdots & C_j & \cdots & C_n \end{array} \\
\begin{bmatrix}
(l_{11(w)}, m_{11(w)}, u_{11(w)}) & \cdots & (l_{1j(w)}, m_{1j(w)}, u_{1j(w)}) & \cdots & (l_{1n(w)}, m_{1n(w)}, u_{1n(w)}) \\
\cdots & \cdots & \cdots & \cdots & \cdots \\
(l_{i1(w)}, m_{i1(w)}, u_{i1(w)}) & \cdots & (l_{ij(w)}, m_{ij(w)}, u_{ij(w)}) & \cdots & (l_{in(w)}, m_{in(w)}, u_{in(w)}) \\
\cdots & \cdots & \cdots & \cdots & \cdots \\
(l_{r1(w)}, m_{r1(w)}, u_{r1(w)}) & \cdots & (l_{rj(w)}, m_{rj(w)}, u_{rj(w)}) & \cdots & (l_{rn(w)}, mr_{rn(w)}, u_{rn(w)})
\end{bmatrix} \end{array}
\tag{13.4}
$$

where $(l_{ij(w)}, m_{ij(w)}, u_{ij(w)})$ denotes the weight in triangular fuzzy number for the criterion C_j and assessed by ith person (expert, user or DM) in the specified five degrees of gradation.

Step 5: Determination of average fuzzy decision matrix: Average fuzzy decision matrix is calculated from the r number of decision matrices as assessed by the experts using the proposed formula.

$$\bar{x}_{ij} = \left(\frac{1}{r}\sum_{i=1}^{r} l_{ij}, \frac{1}{r}\sum_{i=1}^{r} m_{ij}, \frac{1}{r}\sum_{i=1}^{r} u_{ij} \right) = \left(x_{ij(l)}, x_{ij(m)}, x_{ij(u)} \right) \quad (13.5)$$

Where $\bar{x}_{ij}$ denotes the average fuzzy performance rating, l_{ij}, m_{ij}, u_{ij} are lower, middle and upper values of ith fuzzy performance rating, respectively. r is number of experts involved in decision making process, i varies from 1 to r, j is associated with criteria.

Step 6: Determination of average fuzzy weight matrix: Average fuzzy weight matrix is calculated from the weight matrix of order r×n assessed by r experts. The average weight matrix turns to a 1×n order matrix of fuzzy numbers, which can be determined by the following formula.

$$\bar{v}_j = \left(\frac{1}{r}\sum_{i=1}^{r} l_{ij(w)}, \frac{1}{r}\sum_{i=1}^{r} m_{ij(w)}, \frac{1}{r}\sum_{i=1}^{r} u_{ij(w)} \right) = \left(v_{j(l)}, v_{j(m)}, v_{j(u)} \right) \quad (13.6)$$

where $\bar{v}_j$ denotes the average fuzzy weight of jth criteria. $l_{ij(w)}$, $m_{ij(w)}$, $u_{ij(w)}$ are lower, middle, and upper values of the fuzzy weight assessed by *i*th person (expert) involved in decision making process.

Step 7: Normalization of fuzzy performance rating: Based on the desired value, criteria can be divided in three categories, i.e., benefit criteria (higher value is desired), non-benefit criteria (lower numerical value is desired) and target value oriented criteria (neither higher nor lower value is desired instead a certain fixed value is expected). Since nature of the criteria is different, hence unique normalization technique for each category of criteria is used. Following formulae are proposed for performing normalization process.

$$\tilde{r}_{ij} = \left(r_{ij(l)}, r_{ij(m)}, r_{ij(u)} \right) = \begin{cases} \left(x_{ij(l)}, x_{ij(m)}, x_{ij(u)} \right), j \in B & (13.7) \\ \left(x_{ij(l)\max} - x_{ij(l)},\ x_{ij(m)\max} - x_{ij(m)},\ x_{ij(u)\max} - x_{ij(u)} \right), j \in NB & (13.8) \\ 1 - \left| x_{ij(u)\max} - x_{ij(u)} \right|, 1 - \left| x_{ij(m)\max} - x_{ij(m)} \right|, 1 - \left| x_{ij(l)\max} - x_{ij(l)} \right|, j \in T & (13.9) \end{cases}$$

Step 8: Normalization of weight: For the compatibility of weight with the performance rating, the following normalization technique is introduced.

$$\tilde{w}_j = \left(w_{j(l)}, w_{j(m)}, w_{j(u)}\right) = \left(\frac{v_{j(l)}}{v_{j(u)\max}}, \frac{v_{j(m)}}{v_{j(u)\max}}, \frac{v_{j(u)}}{v_{j(u)\max}}\right) \quad (13.10)$$

where $v_{j(u)\max}$ denotes the maximum upper value of all TFNs expressing weight criteria.

Step 9: Performance contribution of individual criterion: The normalization process is carried out in such a way that all the fuzzy performance ratings are now in benefit sense irrespective of nature of criteria. That is why for finding out the individual contribution of criteria in performance measurement is carried out using a common unique formula as proposed follows.

$$z_{ij} = EXP(r_{ij(l)} + w_{j(l)}) + EXP(r_{ij(m)} + w_{j(m)}) + EXP(r_{ij(u)} + w_{j(u)}) \quad (13.11)$$

$$\text{or } z_{ij} = \sum_{k=l}^{u} EXP(r_{ij(k)} + w_{j(k)}) \quad (13.12)$$

Step 10: Performance Index: The performance index is the measurement in performance, ranking and selection. The current investigation advocates the following technique in calculation performance index (PI) by integrating performance contribution of each criterion toward the evaluation of alternatives.

$$PI_i = \sum_{j=1}^{n} \sum_{k=l}^{u} EXP(r_{ij(k)} + w_{j(k)}) \quad (13.13)$$

Here, PI_i denotes performance index of ith alternative. Performance index has its beneficial sense, that is, higher value is desirable. Therefore alternatives are arranged in the descending order of their respective PI value. The alternative with the highest value of performance index is considered the best alternative. The alternative with the least value of performance index is considered as the worst alternative, and so on.

This proposed algorithm is now demonstrated with a suitable example on selection of ergonomic design problem in the next section.

13.3 An Illustrative Example on Ergonomic Design Evaluation

An East Indian manufacturing organization plans to select the best sitting chairs with respect to ergonomic design considerations. After a wide range of survey, the company collects the following information by questionnaires to four potential users in a certain demographical region. The three users have different physiological heights is formed. User 1 is 5', User 2 is 5'6" and User 3 is 6'. The decision making committee selects seven important criteria as follows: C1, Seat Height (SH); C2, Back Rest Size (BRS); C3, Back Rest Angle (BRA); C4, Arm Rest (AR); C5, Material (M); C6, Load Capacity (LC); and C7, Cost (C). All criteria are of qualitative/intangible categories. The criteria have been assessed using five degrees of linguistic variables along with the corresponding abbreviation and triangular fuzzy number as furnished in Table 13.1. Five degrees of linguistic variables, abbreviation along with corresponding TFN for weights are presented in Table 13.2.

All the criteria under consideration are subjective and vague. Linguistic variables are used by the decision makers for the assessment of the performance rating of the alternatives with respect to the criteria. Decision maker 1 assesses the alternatives with respect to criteria in terms of five degree linguistic variables and forms a decision matrix, which is shown in Table 13.3.

If we analyze Table 13.3 where user 1 has judged the seating chair with linguistic variables, many conflicting factors arises for consideration. Here for chair 1 (Alternative 1), the cost, back rest size, back rest angle are near about desired one but the criteria like seat height, load capacity, etc. are not good in respect of physiological application considering the height of the user. Moreover, the properties of chair 1 will provide postural fatigue

Table 13.3 Decision matrix in linguistic variables assessed by user (expert) 1.

By user 1		Criteria						
		C1	C2	C3	C4	C5	C6	C7
Alternatives	A1	VC	NU	NC	VC	A	NC	A
	A2	A	A	A	C	VC	C	VU
	A3	VU	C	VC	A	VU	VC	NC
	A4	C	VC	C	A	C	C	NC

to the user 1. But in case of user 3, chair 1 is not comfortable in respect of load capacity, back rest angle, back rest size but its performance is near to desired one in respect of seat height, arm rest. So same chair can be comfortable in respect to different criteria to different users, as well as that cannot be desired in respect to other criteria for the same users.

Decision matrix in terms of linguistic variable estimated by users 2 and 3 are accommodated in Table 13.4 and Table 13.5, respectively. The linguistic variables in Tables 13.3, 13.4, and 13.5 are now converted into corresponding TFNs.

The importance of the seven criteria has been estimated by the three users using prescribed five degrees of linguistic variables. Weights in linguistic variables estimated by the users P1, P2, and P3 are shown in Table 13.6. The abbreviations bear the same meaning as describe before. Weights in linguistic variables are now converted in terms of triangular fuzzy numbers as per the conversion rules as describe before.

Table 13.4 Decision matrix in terms of linguistic variable estimated by user 2.

Estimated by user 2		**Criteria**						
		C1	**C2**	**C3**	**C4**	**C5**	**C6**	**C7**
Alternatives	A1	A	V C	C	VU	VC	NC	C
	A2	VC	C	VC	C	C	A	NC
	A3	A	A	A	VC	A	C	A
	A4	C	A	A	NC	C	C	NC

The abbreviations bear the meaning as prescribed in Table 13.1.

Table 13.5 Decision matrix in terms of linguistic variable estimated by user 3.

Estimated by user 3		**Criteria**						
		C1	**C2**	**C3**	**C4**	**C5**	**C6**	**C7**
Alternatives	**A1**	**VC**	**NU**	**NC**	**VC**	**A**	**NC**	**A**
	A2	**A**	**A**	**A**	**C**	**VC**	**C**	**VU**
	A3	**VU**	**C**	**VC**	**A**	**VU**	**VC**	**NC**
	A4	**C**	**VC**	**C**	**A**	**C**	**C**	**NC**

The abbreviations bear the meaning as prescribed in Table 13.1.

Table 13.6 Weights in linguistic variables estimated by users.

Experts	C1	C2	C3	C4	C5	C6	C7
P1	EI	VI	LI	I	VI	I	UI
P2	VI	I	EI	UI	I	LI	I
P3	LI	EI	VI	LI	LI	EI	EI

Average fuzzy performance rating and average fuzzy weight matrix are computed as per equations (13.5) and equation (13.6) respectively. Normalized fuzzy performance rating and normalized fuzzy weight are calculated using equations (13.7) to equation (13.9) and equation (13.10). Computation of performance contribution of the criteria is carried out using a common unique equation (13.11) and presented in Table 13.7. Performance indices of each alternative design are calculated using equation (13.13), and the results are shown in Table 13.8. The term performance index is introduced in measuring performance, ranking, and selection.

Table 13.7 Performance contribution of individual criterion.

		Criteria						
		C1	C2	C3	C4	C5	C6	C7
Alternative	A1	16	18	17	12	10.73	8.4	7.7
	A2	15	16	17	9.8	11.47	11	12
	A3	16	20	18	9.1	9.39	13	10
	A4	15	15	16	11	10.73	13	11

Table 13.8 Performance index and ranking order of alternatives.

Alternative		Performance index value	Ranking order
Alternatives	A1	90.94	4
	A2	93.53	2
	A3	95.28	1
	A4	92.32	3

The current investigation advocates the following technique in calculation performance index (PI) by integrating performance contribution of each criterion toward the evaluation of alternatives. Figure 13.1 depicts performance index of alternatives. The alternatives are arranged as per the descending order of their performance indices value. It is clearly seen that A3>A2>A4>A1. According to their value, the third alternative has the

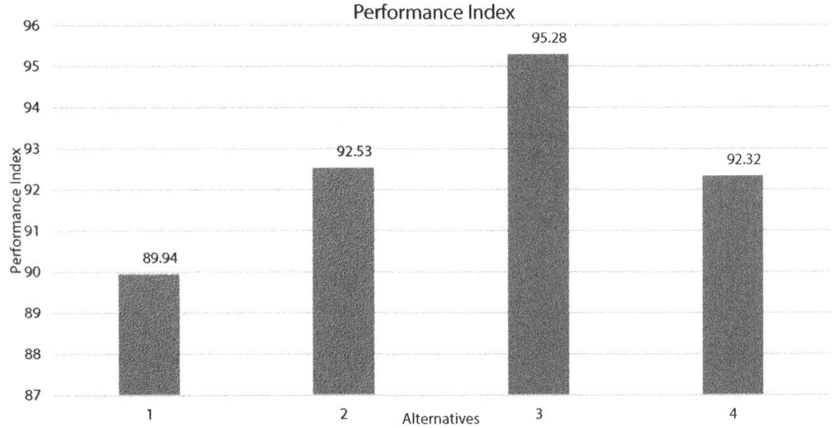

Figure 13.1 Performance index of alternatives.

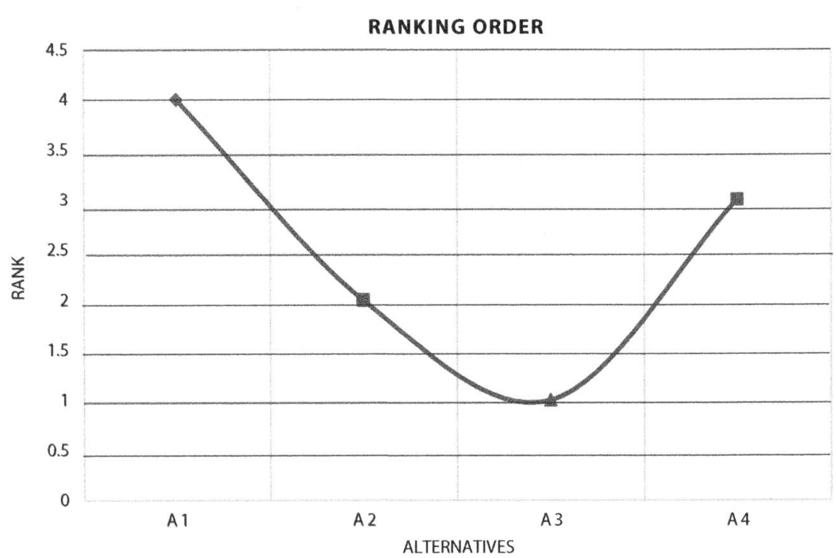

Figure 13.2 Ranking order of the alternatives.

highest score. Calculation shows that A3 is the best alternative of the four, A1 is the worst one, and so on. The ranking order is also graphically shown in Figure 13.2. Comprising all the factors and facts with this new set of algorithm, the selection has been conducted, which can analyze and solve complex problem into one point of perseverance.

13.4 Conclusions

The present state of knowledge of ergonomically design selection is mainly limited by consideration of objective criteria with crisp value. The current investigation explores a technique having the ability of incorporation of target-based criteria, benefit criteria and cost criteria under fuzzy environment with vague information. Thus, this study contributes new concept in the domain of knowledge by introducing an innovative technique capable of judging intangible factors as per the requirement of users in varying social environment. The proposed multicriteria decision making approach is useful and effective in evaluation and selection of the best alternative under fuzzy environment. This technique may be an important aid of the decision makers for selection of the best ergonomic design. Consideration of interdependent criteria involving and heterogeneous decision makers might be direction of future research.

References

1. Jain, R., Rana, K.B., Meena, M.L., An integrated multi-criteria decision-making approach for identifying the risk level of musculoskeletal disorders among handheld device users. *Soft Comput.*, 2021, https://doi.org/10.1007/s00500-021-05592-w.
2. Eraslan, E., Güneşli, İ., Khatib, W., The evaluation of appropriate office layout design with MCDM techniques. *SN Appl. Sci.*, 2388, 2020, https://doi.org/10.1007/s42452-020-2181-x.
3. Mohanty, P.P., Mahapatra, S.S., Mohanty, A., Sthitapragyan, A novel multi-attribute decision making approach for selection of appropriate product conforming ergonomic considerations. *Oper. Res. Perspect.*, 5, 82–93, 2018.
4. Halder, P., Karmaker, C., Rahman, M., *Integrated Approach of Ergonomics and MCDM into Truck Drivers' Seat Comfort: A Case Study in Bangladesh International Conference on Mechanical, Industrial and Energy Engineering 2018*, Khulna Bangladesh, 23-24 December 2018.
5. Mohammad, K., Majid, A., Alireza, K., Saeedeh, M., Integrating the ergonomics techniques with multi criteria decision making as a new approach

for risk management: An assessment of repetitive tasks-entropy case study. *J. Res. Health Sci.*, 16 , 2, 85–89, 2016.
6. Hignett, M.A., 2000REBA (Rapid Entire Body Assessment). *Appl. Ergon.*, 31, 201–205, 2000.
7. Fransson-Hall, C., Gloria, R., Kilbom, A., Winkel, J., A portable ergonomic observation method (PEO) for computerised on-line recording of postures and manual handling. *Appl. Ergon.*, 26, 2, 93–100, 1995.
8. Karhu, O., Kansi, P., Kuorinka, I., Correcting working postures in industry: A practical method for analysis. *Appl. Ergonomics*, 8, 199–201, 1977.
9. Waters, T.R., Putz-Anderson, V., Garg, A., Fine, L.J., Revised NIOSH equation for the design and evaluation of manual lifting tasks. *Ergonomics*, 36, 749–776, 1993.
10. Habibi, E., Asaadi, Z., Hosseini, M., Proportion of elementary school pupils anthropometric characteristics with dimensions of classroom furniture in Isfahan, Iran. *J. Res. Med. Sci.*, 16, 98–104, 2011.
11. Macedo, A., Morais, A.V., Martins, H.F., Martins, I.C., Pais, S.M., Mayan, O.S., Match between classroom dimensions and students anthropometry: Re-equipment according to European educational furniture standard, Human Factors. Int. J. Hum. *Factors Ergon. Soc.*, 57, 48–60, 2015.
12. Castellucci, H.I., Arezes, P.M., Viviani, C.A., Mismatch between classroom furniture and anthropometric measures in Chilean schools, *Appl. Ergon.*, 41, 563–568, 2010.
13. Mohamed-Thariq, M.G., Munasinghe, H.P., Abeysekra, J.D., Designing chairs with mounted desktop for university students: Ergonomics and comfort. *Int. J. Ind. Ergon.*, 40, 8–18, 2010.
14. Xu, J. and Zhang, H., Modern office furniture design based on ergonomics. *Adv. Mater. Res.*, 62, 57–62, 2013.
15. Sepehri, S., Habib, A.H., Shakerian, S., The relationship between ergonomic chair and musculoskeletal disorders in north of Khuzestan's students. *Eur. J. Exp. Biol.*, 3, 181–187, 2013.
16. Shah, R.M., Bhuiyan, M.A.U., Debnath, R., Iqbal, M., Shamsuzzoha, A., Ergonomics issues in furniture design: A case of a tabloid chair, in: *Proc. of the 23rd International Conference on Flexible Automation and Intelligent Manufacturing (FAIM 2013)*, June 26th - 28th, Porto, Portugal, pp. 91–104, 2013.
17. Altaboli, I., Ahmida, R., Elmgrab, M., Immraga, H., Othman, R., Anthropometric evaluation of proposed improved designs of the classroom desk for Benghazi primary schools, in: *Proc. of the Human Factors and Ergonomics Society Annual Meeting*, vol. 59, pp. 426–430, 2015.
18. Tunay, M. and Melemez, K., An analysis of biomechanical and anthropometric parameters on classroom furniture design. *Afr. J. Biotechnol.*, 78, 1081–1086, 2008.
19. Lane, K. and Richardson, M., Classroom furniture and postural alignment. *Educ. Facil. Planner*, 21, 22–23, 1993.

20. Mandal, A., The correct height of school furniture. *Hum. Factors*, 24, 257–269, 1982.
21. Toomingas, I. and Gavhed, D., Workstation layout and work postures at call centres in Sweden in relation to national law, EU-directives and ISO-standards, and to operators comfort and symptoms. *Int. J. Ind. Ergon.*, 38, 1051–1061, 2008.
22. Linton, S., Hellsing, A., Halme, T., The effects of ergonomically designed school furniture on pupils attitudes, symptoms and behavior. *Appl. Ergon.*, 25, 299–304, 1994.

14

Optimization of Energy Generated from Ocean Wave Energy Using Fuzzy Logic

S. B. Goyal[1*], Pradeep Bedi[2], Jugnesh Kumar[3] and Prasenjit Chatterjee[4]

[1]City University, Petaling Jaya, Selangor, Malaysia
[2]Galgotais University, Greater Noida, U.P., India
[3]St. Andrews Institute of Technology & Management, Gurgaon, India
[4]MCKV Institute of Engineering, West Bengal, India

Abstract

The popularity of energy generation from renewable resources had increased in the last decades. Different sources, such as solar energy, wind energy, biogas energy, etc, are the most popular energy sources. But ocean wave energy as an alternative to conventional energy sources is also gaining much attention from researchers. As earth contains approx. 71% of water all around and the ocean contains approx. 97% of the total water on earth. So, this can be the most effective medium to utilize such a huge resource to fulfill the energy requirement of people worldwide. Many factors can determine the amount of power generated, such as wave height, time period, frequency, etc. These factors influence the amount of utilizable wave energy potential. In this chapter, a model is presented that can evaluate the power generated from ocean waves. Further, it can maximize the overall power generated by the application of a fuzzy optimizer.

***Keywords*:** Ocean waves, energy generation, wave energy converter, optimization, fuzzy logic

**Corresponding author*: drsbgoyal@gmail.com

Sachi Nandan Mohanty, Prasenjit Chatterjee and Bui Thanh Hung (eds.) Fuzzy Computing in Data Science: Applications and Challenges, (253–266) © 2023 Scrivener Publishing LLC

14.1 Introduction

Renewable energy resources are the most abundant and clean infinite energy reserves that need to be extracted properly for energy generation to combat environmental degradation. This includes various resources, such as solar, wind, wave energy, etc. Of these resources, the wave energy is available with a higher-energy density that needs to be extracted deploying a proper conversion technology [1]. In recent years, there have been a large study on various wave energy generation systems by different researchers to extract maximum energy from them [2, 3]. Researchers have classified the wave energy generating station into three types, namely single-body floating buoy, two-body floating buoys, and multibody floating buoys by the number of floating buoys [4], as shown in Figure 14.1.

"Girard and Son" a company from France were the first one to study the possibilities of generating power from these resources and have patented their idea in early 1799 [5]. After their idea, the first device employed and its development for wave energy extraction was studied and patented by Leishman and Scobie in 1855 up to 1973. Since then various countries such as United Kingdom, Portugal, Ireland, Norway, Sweden, and Denmark have been searching the possibilities of extracting the energy from these resources efficiently for industrial power consumption in short-term and long-term systems [5]. Figure 14.2 depicts the nature of the ocean waves [6, 7].

Figure 14.1 Types of buoys for ocean wave.

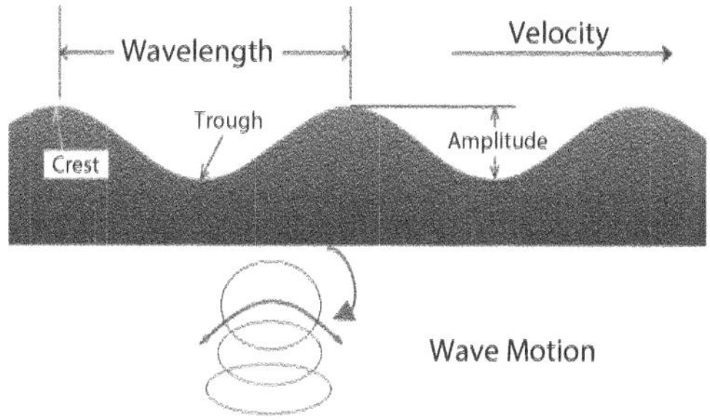

Figure 14.2 Waves motion and depth.

Equation 14.1 shows the basic formula for energy extracted from these systems and was found to be proportional to the square of amplitude and period of the wave in motion.

$$E = \frac{1}{2}\rho g A^2 V_g \tag{14.1}$$

where ρ = Water density
g = Gravitational constant
A = Wave amplitude
Vg = Wave speed.

Even though the waves in the ocean have tremendous energy stored in them, we need an efficient conversion system to extract the maximum of it for industrial and domestic use. The problems in areas in designing the control system include peak-to-average load ratios, which are highly random and varying and the prediction of the loads becomes difficult before implementing the device in the ocean. Thus, the implementation of any wave energy control (WEC) requires a deep study of the wave nature, frequency, and amplitude to make the system economical and favorable. WECs are also advantageous in terms of environmental factors [8].

14.2 Control Approach in Wave Energy Systems

Many conversion systems have been studied by researchers around the world, and one common mechanism was found in all of them, that is, each of them employed a mechanical conversion system to study the wave nature and its variation and an electrical generating station to convert the mechanical energy into electrical form.

The mechanical conversion device converts the wave energy of the tides into a more usable form for which it may employ various hydraulic coupling arrangements, gearboxes, etc. The system may result in added mechanical losses and does require extra maintenance for the device. Therefore, researchers have been trying to focus on making use of generators associated with the sea waves to reduce these extra losses and maintenance costs. Therefore, linear generators are being studied for the extraction of wave power from the oceans to a more usable form [9, 10]. For the extraction of maximum power/electricity, there is a need to consider the mathematical evaluation of the energy features needed for supply. The magnetic design plays a key role in influencing the energy outcomes from the conversion system. In the study to combat such difficulties of the systems, a permanent magnet (PM) linear electrical generator design has been made and built to attain effectiveness. Figure 14.3 shows the basic converter technology being used in wave energy generating stations.

The converters in the wave energy systems make use of different technologies which is being divided into three main categories as follows [8]:

Oscillating water column (OWC)
Oscillating body converter (OBC)
Overtopping converters (OTC)

An oscillating water column is the kind of WEC, which are meant to absorb the power resulting from the oscillation of seawater inside a chamber structure resulting from the waves. This chamber can be a constant one or a floating shape that merely depends on the convention. The oscillating body converter is the converter that moves along with the wave oscillation.

These devices can be floating on the waves or fully submerged and hence are capable of extracting more energy from the strong waves in the oceans.

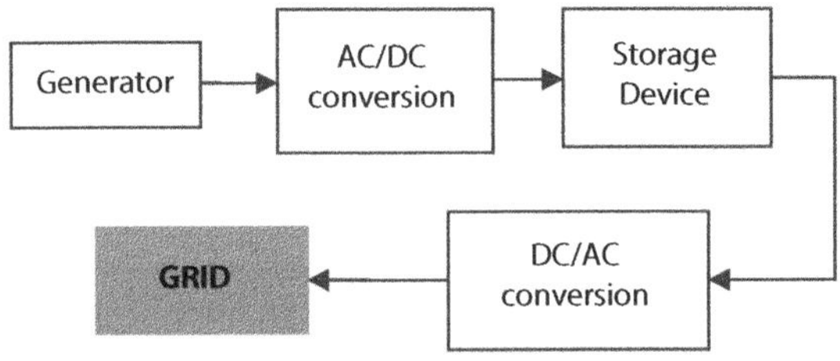

Figure 14.3 Block diagram of a sea wave energy harvesting device.

CorPower Ocean AB is one of those institutions that develop this floating type oscillating body converter.

Due to their area of installation, these devices are also addressed as off-shore or as third-generation devices and found to be more complex when we compared them with OWC. Overtopping converters elevates are the converters that elevate the waves of the ocean above the sea level, and the water is then drawn from a height to the turbine and then to the generator for power generation in the form of electricity.

14.3 Related Work

Falnes *et al.* [1] discussed the wave spectrum parameters associated with shipping, distribution, and variability of wave energy within the sea for energy extraction. Many unique kinds of wave-energy converters, of numerous categories, had been brought into study. They have considered the primary conversion of wave mechanical energy by using an oscillating device as a wave-interference phenomenon and thus, corresponding to optimum wave interference have proposed that there is a higher bound to the quantity of energy that is possible to be extracted from a wave by a selected oscillating system. They have derived an upper bound for the ration of the wave energy extracted to the volume of the wave energy immersed in the oceans. In the end, the two upper bounds and their significance were also studied in this review paper. Falcao *et al.* [2] focused on the various factors associated with the development of wave energy electricity since 1970. Various topics have been brought into studies such as wave power as a resource, briefing of its historical background in respect of hydrodynamics of wave energy absorption, and its controlling. The paper discusses the various studies carried following the wave energy conversion devices and its development in recent years. The concept, design, version-testing, creation, and deployment into the real sea of prototypes; and the improvement of sea equipment (air and water generators, excessive-stress hydraulics, linear electric turbines) and mooring structures.

Chen *et al.* [4] focused on the extraction of energy from offshore waves in the ocean since it has been found that extraction of energy can be easily predicted in offshore waves as compared to other systems such as shoreline waves, near-shore waves, and tides. The paper has focused on various aspects of this generating station, such as wave energy extraction technologies and the linear generator and wave interaction. The study has depicted that the design of floating buoy and linear generators performs a critical function within the operational performance of offshore wave energy extraction machine. This conclusion is meant to define paths for further researches to be done in this field.

Sešek *et al.* [5] focused their work on the erosion done by the sea waves, which may result in deterioration of the infrastructure and unwanted shore changes, which may finally result in floods and pollution. My device is being proposed by the paper for monitoring of wave movement and analysis with the basic sensor and its result is also measured and projected.

Ringwood *et al.* [11] presented an overview of the control strategy of energy maximization techniques of wave energy converters. In this paper, the researcher explained the mathematical model and two control strategies along with algorithms for forecasting of eave energy generated. Out of two, one of the control strategies is optimal control strategy.

Feng *et al.* [12] analyzed the three, five, and eight float wave energy converter arrays. Researchers have implemented a differential evolutionary algorithm to achieve fast conversion and accurate results. The optimal results are obtained with an eight-float setup. It was observed that the greater the interaction between wave energy and converter, the higher the efficiency.

Neshat *et al.* [13] explored the optimization of energy generated from a three-buoy wave energy converter. The total energy of the system is optimized by adjusting the position of buoys. This system takes wave parameters on the optimal location of buoys and generates maximum power. The algorithm is named as hybrid cooperative coevolution algorithm having local search strategy and backtracking strategy for optimizing the positions of WECs, respectively.

Neshat *et al.* [14] proposed the optimization of a wave energy converter and placement of oscillating buoy. In this work, research proposed the design of wave farms with three-buoy architecture that is fully submerged and oscillating. The design is focused on power maximization and extract maximum output by placing buoys at optimal places. The result was evaluated on numerical optimization and calculation.

The increase in the electricity demand of the population can be efficiently tackled by improving the control system in the wave energy extraction infrastructure. India being a peninsular country is surrounded by the oceans on three sides and is exposed to enormous wave energy reservoirs. The enhancement in the design of the wave energy trappers along with the controller for the electronic converters and reduce the overall cost of the generating system. The area deploys a wide area for R/D in predicting the waves and loading parameters at the site to enhance the power output from the system and its smooth operation.

14.4 Mathematical Modeling for Energy Conversion from Ocean Waves

Most of the wave energy converter (WECs) extracts energy from an ocean wave. These systems don't store energy and termed as a real wave conversion process which means if the wave is strong then the energy generated will be more. Otherwise, energy generated will be less in the condition of weak waves. Sometimes, it may not generate energy [14]. Ocean wave converter (OWC) uses constant pressure to extract and store energy. The pressure is used to rotate the turbine and generate electricity. Figure 14.4 illustrates the block diagram of the conceptual model for electricity generation. Ocean energy is calculated as a summation of potential energy (PE) and kinetic energy (KE).

The wave energy is considered as sine wave having up and down motion results in vibration, represented as simple harmonic motion (SHM) as expressed below:

$$y = A sin \frac{2\pi}{\lambda}(vt - x) \tag{14.2}$$

where:

A= Sea Wave height (in m)
v: Wave propagation velocity (in m/s)
λ: Wave length (in m)
t: Wave peak time (in s)

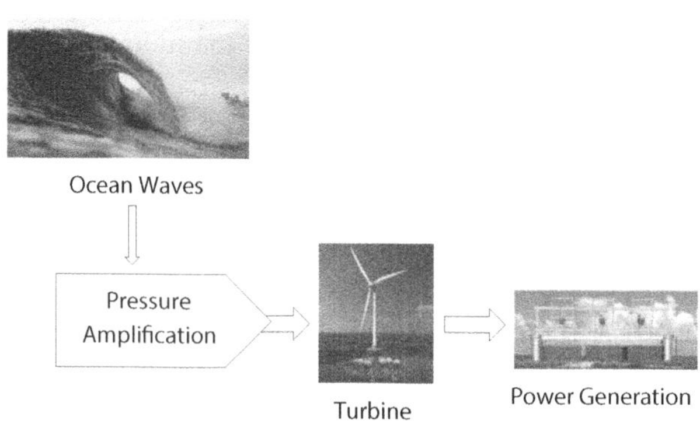

Figure 14.4 Conceptual model wave energy generation.

The particle velocity vp, is calculated as:

$$v_p = \frac{dy}{dt} = \frac{2\pi A v}{\lambda} cos \frac{2\pi}{\lambda}(vt - x) \tag{14.3}$$

Potential energy is calculated as:

$$PE = \frac{2\pi^2 \rho v^2}{\lambda^2} A^2 sin^2\left(\frac{2\pi}{\lambda}(vt - x)\right) \tag{14.4}$$

The kinetic energy is calculated as:

$$KE = \frac{1}{2}\rho v_p{}^2 = \frac{2\pi^2 \rho^2}{\lambda^2} A^2 cos^2\left(\frac{2\pi}{\lambda}(vt - x)\right) \tag{14.5}$$

The total energy generated per unit volume is calculated as:

$$E_t = PE + KE = \frac{2\pi^2 \rho v^2}{\lambda^2} A^2 \tag{14.6}$$

14.5 Proposed Methodology

The following problems are faced: The power generated from solar energy is not constant throughout the year. It is weather dependent and its efficiency drops due to weather conditions. The maintenance cost of solar as well as wind energy is high.

Existing wave energy converters (WECs) do not use an energy storage system and depend on real-time extraction which means when strong waves are generated then only energy is generated. Need to focus on other ocean parameters other than wave height and to optimize them for maximization of output power.

In this work, a model is presented that can consider different wave parameters that can be optimized to maximize the power generated from wave energy using fuzzy logic. As power generated from ocean waves may

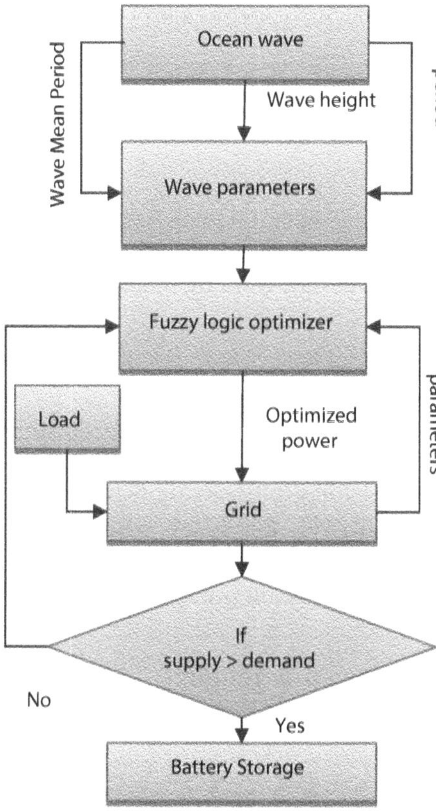

Figure 14.5 Proposed wave optimization model.

be low as compared to other resources. So, it is needed to optimize the available resource to its maximum level. Different ocean parameters are fed into the fuzzy that will result in maximum output from the same input. The proposed flow diagram is shown in Figure 14.5.

14.5.1 Wave Parameters

As mentioned above, some wave parameters that are needed to be optimized to extract maximum power:

Wave Height: The wave height is one of the most important parameters that is used to determine the wave energy. It is calculated as the difference between ocean wave trough and crest, as ocean wave is similar to a sine wave.

Peak Period: The peak period is defined as:

$$T_p = \frac{1}{f_p} \tag{14.7}$$

where f_p = maximum value of spectral peak frequency.

Wave Mean Period: The wave mean period is calculated as the average time period that is used to design wave turbines for energy generation.

14.5.2 Fuzzy-Optimizer

The procedure of the Fuzzy-Optimizer method is as follows:

Step 1: Assignment of decision group having N members.
Step 2: Concerning input, X_{ij}, the fuzzy output is generated with n decision-makers.
Step 3: Evaluation of aggregate result out of decisions.

$$a_{ij} = \min\{a_{ij}^n\} \tag{14.8}$$

$$b_{ij} = \frac{1}{N}\sum_{n=1}^{N} b_{ij}^n \tag{14.9}$$

$$c_{ij} = \min\{c_{ij}^n\} \tag{14.10}$$

where X_{ij} = (a_{ij}, b_{ij}, and c_{ij}) are decision makers having weight w_{ij} = (w_{ij1}, w_{ij2}, w_{ij3}) calculated as:

$$w_{ij1} = \min\{w_{ij1}^n\} \tag{14.11}$$

$$w_{ij2} = \frac{1}{N}\sum_{n=1}^{N} w_{ij2}^n \tag{14.12}$$

$$w_{ij3} = \min\{w_{ij3}^n\} \tag{14.13}$$

Step 4: Compute Normalized fuzzy decision matrix

$$r_{ij} = \left(\frac{a_{ij}}{c_j}, \frac{b_{ij}}{c_j}, \frac{c_{ij}}{c_j} \right) \text{ and } c_j = \max\{c_{ij}^n\} \tag{14.14}$$

$$r_{ij} = \left(\frac{a_j}{c_j}, \frac{b_j}{c_j}, \frac{c_j}{c_j} \right) \text{ and } c_j = \max\{a_{ij}^n\} \tag{14.15}$$

$$v_{ij} = r_{ij} * w_{ij} \tag{14.16}$$

Step 5: Compute the fuzzy true (positive) solution and fuzzy false (negative) solution. The FTS and FFS are calculated as follows (Table 14.1):

$$FTS = \max(v_{ij}) \tag{14.17}$$

$$FFS = \min(v_{ij}) \tag{14.18}$$

The proposed optimizer accepts three input variables related to waves, such as wave height, Peak period, and mean period. The fuzzy optimizer outputs two crisp variables namely, the pressure amplification value and torque. This amplified pressure is used to give high torque to rotate the turbine system for power generation. in grid Seven variables are assigned for

Table 14.1 Fuzzy rules for optimization of parameters.

	FB	FM	FS	Z	TS	TM	TB
FB	FB	FB	FM	FM	FS	FS	Z
FM	FB	FM	FM	FS	FS	Z	TS
FS	FM	FM	FS	FS	Z	TS	TS
Z	FM	FS	FS	Z	TS	TS	TM
TS	FS	FS	Z	TS	TS	TM	TM
TM	FS	Z	TS	TS	TM	TM	TB
TB	Z	TS	TS	TM	TM	TB	TB

each input and output variable, that is FB (false big), FM (false medium), FS (false small), Z (zero), TS (true small), TM (true medium), and TB (true big). A sample fuzzy rule table is given in Table 14.1. The membership function will be chosen for all input and output variables in the specified range.

14.6 Conclusion

The primary focus of this work is to use other renewable resources, such as ocean energy to generate power, which will give a subject for future research. In this paper, a model is presented using fuzzy logic control whose function is to optimize energy generated by ocean waves. These waves are first tracked, and then its influencing factors are evaluated whose variation impact generated power. The fuzzy optimizer is designed on fuzzy membership functions on three variables, such as wave height, wave peak period time, and wave mean height produced. These wave parameters will decide the frequency of waves generated and the power extracted. The generated power is supplied to the grid whose load is fluctuating in nature. So, these load parameters are also fed into a fuzzy system to optimize energy according to demand. If there is a shortage, then the system will generate an alarm that will help to compensate for energy shortage. Therefore, to meet the future demand for renewable energy sources, ocean wave energy is one of the effective medium for researchers.

References

1. Falnes, J., A review of wave-energy extraction. *Mar. Struct.*, 20, 185–201, 2007. https://doi.org/10.1016/J.MARSTRUC.2007.09.001.
2. Falcão, A.F.D.O., Wave energy utilization: A review of the technologies. *Renew. Sustain. Energy Rev.*, 14, 899–918, 2010, https://doi.org/10.1016/J.RSER.2009.11.003.
3. Oğuz, Y., Güney, I., Çalik, H., Power quality control and design of power converter for variable-speed wind energy conversion system with permanent-magnet synchronous generator. *Sci. World J.*, 2013, 1–14, 2013, https://doi.org/10.1155/2013/783010.
4. Chen, Z., Yu, H., Hu, M., Meng, G., Wen, C., A review of offshore wave energy extraction system. *Adv. Mech. Eng.*, 623020, 1-9, 2013.
5. Sešek and Trontelj, J., *Measurement System for Sea Wave Monitoring*, Procs. of 49th International Conference on Microelectronics, Devices and Materials, Kranjska Gora, Slovenia, September 25th - 27th, 2013.

6. Clément, A., McCullen, P., Falcão, A., Fiorentino, A., Gardner, F., Hammarlund, K., Lemonis, G., Lewis, T., Nielsen, K., Petroncini, S., Pontes, M.T., Schild, P., Sjöström, B.O., Sørensen, H.C., Thorpe, T., Wave energy in Europe: Current status and perspectives. *Renew. Sustain. Energy Rev.*, 6, 405–431, 2002, https://doi.org/10.1016/S1364-0321(02)00009-6.
7. Leijon, M., Bernhoff, H., Ågren, O., Isberg, J., Sundberg, J., Berg, M., Karlsson, K.E., Wolfbrandt, A., Multiphysics simulation of wave energy to electric energy conversion by permanent magnet linear generator. *IEEE Trans. Energy Convers.*, 20, 219–224, 2005, https://doi.org/10.1109/TEC.2004.827709.
8. Polinder, H., Damen, M.E.C., Gardner, F., Linear PM generator system for wave energy conversion in the AWS. *IEEE Trans. Energy Convers.*, 19, 583–589, 2004, https://doi.org/10.1109/TEC.2004.827717.
9. Trapanese, M., Optimization of a sea wave energy harvesting electromagnetic device. *IEEE Trans. Magn.*, 44, 4365–4368, 2008, https://doi.org/10.1109/TMAG.2008.2002192.
10. Anderlini, E., Forehand, D.I.M., Bannon, E., Abusara, M., Reactive control of a wave energy converter using artificial neural networks. *Int. J. Mar. Energy*, 19, 207–220, 2017, https://doi.org/10.1016/J.IJOME.2017.08.001.
11. Ringwood, J.V., Bacelli, G., Fusco, F., Control, forecasting and optimisation for wave energy conversion. *IFAC Proc.*, 47, 7678–7689, 2014, https://doi.org/10.3182/20140824-6-ZA-1003.00517.
12. Fang, H.W., Feng, Y.Z., Li, G.P., Optimization of wave energy converter arrays by an improved differential evolution algorithm. *Energies*, 11, 3522, 2018, https://doi.org/10.3390/EN11123522.
13. Neshat, M., Alexander, B., Wagner, M., A hybrid cooperative co-evolution algorithm framework for optimising power take off and placements of wave energy converters. *Inf. Sci.* (Ny), 534, 218–244, 2020, https://doi.org/10.1016/J.INS.2020.03.112.
14. Neshat, M., Alexander, B., Sergiienko, N.Y., Wagner, M., New insights into position optimisation of wave energy converters using hybrid local search. *Swarm Evol. Comput.*, 59, 100744, 2020, https://doi.org/10.1016/J.SWEVO.2020.100744.

15

The m-Polar Fuzzy TOPSIS Method for NTM Selection

Madan Jagtap* **and Prasad Karande**

Mechanical Engineering Department, Veermata Jijabai Technological Institute, Mumbai, India

Abstract

The m-polar fuzzy set hybrid methods are the subjects of research. The applications of the m-polar fuzzy set hybrid methods are explored in different fields by researchers. One of the m-polar fuzzy set hybrid methodologies is m-polar fuzzy TOPSIS (a technique for order performance by similarity to ideal solution), used in industrial selection problems. In the paper, the nontraditional machining process (NTM) selection problem is solved with the m-polar fuzzy set TOPSIS method. Two criteria weight calculation methods are used while implementing the m-polar fuzzy set TOPSIS method; analytical hierarchy process (AHP) and Shannon's entropy weight calculation methods are used. Effect of input data variation is analyzed on the Euclidean distance of alternatives from a positive ideal solution and Euclidean distance of alternatives from negative ideal solution for AHP weights and Shannon's entropy weight, referred to as positive uncertainty and negative uncertainty. In this paper, results obtained with both ways are compared for rank performance. It can be concluded that there is no effect due to the change in the criteria weight calculation method on rank performance.

***Keywords*:** TOPSIS, m-polar, MCDM, MCGDM, NTM

**Corresponding author*: madan.jagtap@sce.edu.in

Sachi Nandan Mohanty, Prasenjit Chatterjee and Bui Thanh Hung (eds.) Fuzzy Computing in Data Science: Applications and Challenges, (267–286) © 2023 Scrivener Publishing LLC

15.1 Introduction

The m-polar fuzzy set (mFS) algorithm was developed by Chen [1]. It solves multi-criteria group decision making (MCGDM) problems. The mFS is integrated with different multi-criteria decision making (MCDM) methods to form mFS hybrid methods. Hybrid approaches are developed and implemented by Akram [2, 3]. Some of the m-polar fuzzy set hybrid methods are mFS ELECTRE-I (elimination and et choice translating reality), mFS TOPSIS (a technique for order performance by similarity to ideal solution), etc. The mFS ELECTRE-I algorithm is implemented to single-valued input and validated for robot selection problems [16]. Researchers have developed their algorithms and implemented them to real-life issues from social and medical domains. This study's motivation is to explore further these methods for engineering applications with slight changes in the algorithm. Criteria weight calculation methods are added for solving engineering applications. Analytical hierarchy process (AHP) and Shannon's entropy weight calculation methods are used. This paper is divided into six headings. A literature survey follows the first heading in the introduction. The third heading consists of an explanation of methodology. The fourth heading is an industrial case study with its implementation. The non-traditional machining (NTM) process selection problem is considered for the performance of the mFS TOPSIS method. The fifth heading is the results and discussion. The final sixth heading is the conclusion and future scope.

15.2 Literature Review

The mFS TOPSIS (technique for order preference by similarity to ideal solution) methodology helps solve problems with the hesitant fuzzy set, applied to selecting a perfect brand name and selecting a suitable product design for a company [2]. A model for decision-making based on quality function deployment (QFD) in NTM processes selection was developed by Prasad [4]. Chakladar [5] solved four examples of NTM process selection with a diagraph based expert system. Preference ranking organization method for enrichment evaluation (PROMETHEE)-geometrical analysis for interactive aid (GAIA) method was studied and applied for selection of process, which gives a visual decision aid to decision managers [6]. Organization, Rangement Et Synthese De Donnes Relationnelles (ORESTE) method is helpful to analyze various industrial selection problems, which shows that

Table 15.1 Comparison of different NTM process selection methods [11].

Sr. no.	Name of method	Material and shape feature combination			
		Aluminium-surface of revolution	**Ceramics-cylindrical through holes**	**Titanium-deep through cutting**	**Aluminium-precision small holes**
1	AHP-based approach	ECM–PAM	USM–AJM–EBM–LBM	PAM–ECM–AJM	EBM–PAM–LBM
2	Diagraph-based approach	ECM–PAM	USM–AJM–EBM	PAM–EBM–AJM	EBM–AJM–PAM–CHM–LBM
3	ANP-based approach	ECM–PAM	USM–WJM–LBM–AJM–EBM	PAM–ECM–WEDM–WJM	WEDM–CHM–LBM–EBM

it can effectively handle ordinal data for a given situation [7]. TOPSIS-AHP combined approach was used to solve the NTM process selection for the surface of revolution feature generation in stainless steel, also for efficient machining of precision holes on Duralumin (aluminium alloy), cylindrical through-holes generation on ceramic (non-conductive), and standard through-holes generation on Titanium [8]. The multi-objective optimization based on ratio analysis (MOORA) method implemented for solving complex problems of NTM process selection, MOORA method approach is robust and straightforward [9]. TOPSIS-AHP expert system developed by Choudhury [10] works efficiently for selecting a suitable NTM process in the surface of revolution on aluminium, considering a wide variety of attributes. Analytical network process (ANP) shows satisfactory results for setting appropriate NTM process with different materials and shape features as shown in Table 15.1.

Roy [12] has combined fuzzy-AHP and QFD techniques for selecting NTM process, fuzzy-AHP compared the importance of various NTM processes for product and process characteristics, and QFD gives final scores considering shape feature and material combination of NTM process. The Data envelopment analysis (DEA) assisted by multi-attribute decision making (MADM) method implemented to efficiently select suitable NTM processes for various shape features [13]. WASPAS (Weighted aggregated sum product assessment) and COPRAS (complex proportional evaluation assessment) methods were implemented in the NTM process selection for ceramic machining and found helpful in the application [14]. The case-based reasoning (CBR) approach was developed to select the NTM process, software was produced, and three different problems were cited to explain the potential of the CBR approach [15]. Variation in input data results in uncertainty is described by the framework [17]. Effect of normalization on mFS ELECTRE-I method was studied and found vector normalization as suitable normalization [18].

15.3 Methodology

15.3.1 Steps of the mFS TOPSIS

Let, A be the set of alternatives and V be the set of variables,

1. Set of alternatives

$$A = (a_1, a_2, a_3, \ldots\ldots\ldots a_p),$$

2. Set of variablcs

$$V = \{v_1, v_2, v_3, \ldots\ldots\ldots v_q),$$

3. Let, D be set of alternatives and variables, Decision Matrix

$$D = \{ d_{jk}^{\,1}, d_{jk}^{\,2}, d_{jk}^{\,3}, \ldots\ldots d_{jk}^{\,m}\} \text{ where } d_{jk}^{\,i} = d_{jk}(a_j, v_k)$$

$$j = 1,2,3\ldots\ldots p\,,\ k = 1,2,3\ldots\ldots.q \text{ and } i = 1,2,3\ldots\ldots\ldots m$$

4. W is the set of all the criteria weights obtained by either the AHP method or Shannon's entropy weight calculation method,

Weights of criteria $W = (w_1, w_2, w_3, \ldots\ldots w_k)$ where $k = 1,2,3 \ldots\ldots.q,$

5. F, is the weight multiplied decision matrix, Weight multiplied decision matrix

$$F = [(f_{jk}^{\,1}, f_{jk}^{\,2}, f_{jk}^{\,3}, \ldots\ldots\ldots f_{jk}^{\,m})]$$

$$\text{Where } f_{jk}^{\,1} = w_1\, d_{jk}^{\,1} \ldots\ldots\ldots f_{jk}^{\,m} = w_k\, d_{jk}^{\,m}$$

6. The mF positive ideal solution and the mF negative ideal solution of alternatives, mP_{IS} and mN_{IS} are the positive ideal solution and negative ideal solution of alternatives

$$mP_{IS} = \{((d_1^1)^+, (d_1^2)^+,..,(d_1^m)^+),((d_2^1)^+, (d_2^2)^+,...,(d_2^m)^+),..,((d_q^1)^+, (d_q^2)^+,...,(d_q^m)^+)\} \tag{15.1}$$

$$mN_{IS} = \{((d_1^1)^-, (d_1^2)^-,..,(d_1^m)^-),((d_2^1)^-, (d_2^2)^-,...,(d_2^m)^-),...,((d_q^1)^-, (d_q^2)^-,...,(d_q^m)^-)\} \tag{15.2}$$

Where,

$$(d_k^1)^+ = \max\{f_{jk}^1\},\ (d_k^2)^+ = \max\{f_{jk}^2\}\ldots(d_k^m)^+ = \max\{f_{jk}^m\}$$

$$(d_k^1)^- = \max\{f_{jk}^1\},\ (d_k^2)^- = \max\{f_{jk}^2\}\ldots(d_k^m)^- = \max\{f_{jk}^m\}$$

7. Now, $D_e(a_j,P_{IS})$ be the positive euclidean distances of each alternative and $D_e(a_j,N_{IS})$ be the negative Euclidean distances of each option, which can be calculated as,

$$D_e(a_j,P_{IS}) = \sqrt{\frac{1}{m}\sum_{k=1}^{q}\left[(f_{jk}^{1}-(d_j^{1})^{+})^2+(f_{jk}^{2}-(d_j^{2})^{+})^2+\ldots+(f_{jk}^{m}-(d_j^{m})^{+})^2\right]} \tag{15.3}$$

$$D_e(a_j,N_{IS}) = \sqrt{\frac{1}{m}\sum_{k=1}^{q}\left[(f_{jk}^{1}-(d_j^{1})^{-})^2+(f_{jk}^{2}-(d_j^{2})^{-})^2+\ldots+(f_{jk}^{m}-(d_j^{m})^{-})^2\right]}, \tag{15.4}$$

8. The relative closeness-coefficient (RCC) of alternative can be computed as E_j be the RCC for option.

$$E_j = \frac{D_e(a_j,N_{IS})}{D_e(a_j,P_{IS})+D_e(a_j,N_{IS})}, j = 1,2,3\ldots..,p. \tag{15.5}$$

Alternatives with the highest RCC are considered the best alternative.

15.4 Case Study

To perform deep through cutting operations on titanium materials, eight NTM processes are available. Abrasive jet machining (AJM), ultrasonic machining (USM), chemical machining (CHM), electron beam machining (EBM), laser beam machining (LBM), electrochemical machining (ECM), electrical discharge machining (EDM), and plasma arc machining (PAM) are the processes available as alternatives. Two types of criteria are considered for the selection of a suitable NTM process, quantitative measures such as Tolerance and surface finish (TSF), Material removal rate (MRR), and Power requirement (PR). In contrast, the qualitative criteria are Cost (C), Shape factor (F), and work material type (M). A total of six criteria were considered for the selection process. The analytical hierarchy process (AHP) and Shannon's weight calculation method are two different standards weight calculation methods. The analytical hierarchy process (AHP) is subjective, considering the expert decision for criteria weights. Shannon's entropy weight calculation method uses data available from experiments for criteria weight calculations. The current case study will simultaneously implement the criteria weight calculation methods to analyse their effect on the m-polar fuzzy TOPSIS method for NTM process selection.

15.4.1 Effect of Analytical Hierarchy Process (AHP) Weight Calculation on the mFS TOPSIS Method

The analytical hierarchy process (AHP) weight calculation method is implemented to evaluate criteria weight for the decision matrix shown in Table 15.2. Below Table 15.2 shows readings for different criteria against the respective alternative NTM process.

Table 15.3 shows the normalized decision matrix obtained with the max linear approach. Again, the DM from Table 15.2 is considered for normalization purposes in Table 15.3.

AHP weight calculation performed for criteria weights, Table 15.4 shown below presents pairwise comparison of criteria.

The procedure for AHP weight calculation is done in Table 15.4. Table 15.5 below shows the rank of variables as per weight calculations.

Table 15.2 Decision matrix (DM), Example from [5].

Sr. no.	NTM processes	TSF	MRR	PR	C	F	M
A_1	AJM	2.5	0.8	0.22	1	1	4
A_2	USM	1.0	300	2.4	2	1	4
A_3	CHM	3.0	15	0.4	3	1	4
A_4	EBM	2.5	1.6	0.2	4	4	4
A_5	LBM	2.0	0.1	1.4	3	4	4
A_6	ECM	3.0	1500	100	5	5	4
A_7	EDM	3.5	800	2.7	3	1	5
A_8	PAM	5.0	75000	50	1	5	4

Table 15.3 Normalized DM with linear max N1 method [16].

NTM	TSF	MRR	PR	C	F	M
A_1	0.5	1.07E-05	0.0022	0.2	0.2	0.8
A_2	0.2	0.004	0.024	0.4	0.2	0.8
A_3	0.6	0.0002	0.004	0.6	0.2	0.8
A_4	0.5	2.13E-05	0.002	0.8	0.8	0.8
A_5	0.4	1.33E-06	0.014	0.6	0.8	0.8
A_6	0.6	0.02	1	1	1	0.8
A_7	0.7	0.010667	0.027	0.6	0.2	1
A_8	1	1	0.5	0.2	1	0.8

Table 15.4 Pairwise comparison matrix for criteria.

	TSF	MRR	PR	C	F	M
TSF	1	1/3	2	1/2	1/4	1/5
MRR	3	1	3	2	1/3	1/2
PR	1/2	1/3	1	1/2	1/4	1/5
C	2	1/2	2	1	1/3	1/4
F	4	3	4	3	1	1/2
M	5	2	5	4	2	1

Table 15.5 The rank of criteria based on weight their weights.

Sr. no.	Variable	Priority	Rank
1	TSF	0.0665	5
2	MRR	0.157614	3
3	PR	0.052616	6
4	C	0.095786	4
5	F	0.272725	2
6	M	0.354759	1

Consistency Ratio = 0.02949 and Principal Eigen Vector = 6.185. from the consistency ratio value, we can use the weights obtained from the AHP procedure for further steps. Table 15.6 is the weight multiplied matrix obtained by multiplication of AHP weights to normalization matrix shown in Table 15.3. For the remaining steps of the m-polar fuzzy TOPSIS algorithm, a weight multiplied matrix will be used.

Table 15.6 Weight multiplied matrix.

NTM processes	TSF	MRR	PR	C	F	M
A_1	0.03325	1.68632e-06	0.00011572	0.01914	0.05454	0.28376
A_2	0.0133	0.0006304	0.0012624	0.03828	0.05454	0.28376
A_3	0.0399	3.152e-05	0.0002104	0.05742	0.05454	0.28376
A_4	0.03325	3.35688e-06	0.0001052	0.07656	0.21816	0.28376
A_5	0.0266	2.09608e-07	0.0007364	0.05742	0.21816	0.28376
A_6	0.0399	0.003152	0.0526	0.0957	0.2727	0.28376
A_7	0.04655	0.00168112	0.0014202	0.05742	0.05454	0.3547
A_8	0.0665	0.1576	0.0263	0.01914	0.2727	0.28376

Table 15.7 Euclidean distance of alternatives from a positive ideal solution and negative ideal solution.

Sr. no.	Euclidean distance of alternatives from P_{IS}	Euclidean distances of alternatives from N_{IS}
1	$D_e(A_1, P_{IS}) = 0.295271$	$D_e(A_1, N_{IS}) = 0.01995$
2	$D_e(A_2, P_{IS}) = 0.293306$	$D_e(A_2, N_{IS}) = 0.0191853$
3	$D_e(A_3, P_{IS}) = 0.287005$	$D_e(A_3, N_{IS}) = 0.0466147$
4	$D_e(A_4, P_{IS}) = 0.192539$	$D_e(A_4, N_{IS}) = 0.174547$
5	$D_e(A_5, P_{IS}) = 0.196448$	$D_e(A_5, N_{IS}) = 0.168565$
6	$D_e(A_6, P_{IS}) = 0.17203$	$D_e(A_6, N_{IS}) = 0.238597$
7	$D_e(A_7, P_{IS}) = 0.276388$	$D_e(A_7, N_{IS}) = 0.0872234$
8	$D_e(A_8, P_{IS}) = 0.107636$	$D_e(A_8, N_{IS}) = 0.275586$

The positive ideal solution for the alternatives are shown below calculated with the help of equation 15.1,

$$P_{IS} = \{(0.0665),(0.1576),(0.0526),(0.0957),(0.2727),(0.3547)\}$$

The negative ideal solution for the alternative is shown below calculated with the help of equation 15.2,

$$N_{IS} = \{(0.0133),(2.09608e\text{-}07),(0.0001052),(0.01914),(0.05454),(0.28376)\}$$

Table 15.7, shown below, represents Euclidean distance for alternatives from a positive ideal solution and Euclidean distance for other options from the negative ideal solution obtained from equations (15.3) and (15.4), respectively.

Table 15.8 below shows the final ranking obtained from the RCC of each alternative, the highest the RCC higher will be the rank of other options.

Table 15.8 The rank of NTM processes using the m-polar fuzzy TOPSIS algorithm.

Sr. no.	NTM process	Distance of alternatives from positive ideal solution with AHP weight	Distance of alternatives from negative ideal solution with AHP weight	Rank
1	AJM	0.295271	0.01995	Rank 7: E1,0.0632889
2	USM	0.293306	0.0191853	Rank 8: E2,0.0613946
3	CHM	0.287005	0.0466147	Rank 6: E3,0.139724
4	EBM	0.192539	0.174547	Rank 3: E4,0.475493
5	LBM	0.196448	0.168565	Rank 4: E5,0.461805
6	ECM	0.17203	0.238597	Rank 2: E6,0.581055
7	EDM	0.276388	0.0872234	Rank 5: E7,0.239881
8	PAM	0.107636	0.275586	Rank 1: E8,0.719128

Figure 15.1, shown below, is the graphical representation of rank performance obtained by implementing m-polar fuzzy TOPSIS and AHP weight calculation method for selection of NTM process.

15.4.2 Effect of Shannon's Entropy Weight Calculation on the m-Polar Fuzzy Set TOPSIS Method

Shannon's entropy weight calculation method is implemented to evaluate criteria weight for the decision matrix shown in Table 15.1. Table 15.1 shows readings for different criteria against the respective alternative NTM process.

Table 15.10 shows the normalized DM obtained with the max linear approach. In addition, the DM from Table 15.9 is considered for normalization purposes in Table 15.10.

Shannon's entropy weight calculation method is applied on decision matrix shown in Table 15.9, weight obtained from Shannon's entropy weight calculation are as W_{TSF} = 0.153523, W_{MRR} = 0.200184, W_{PR} = 0.182758,

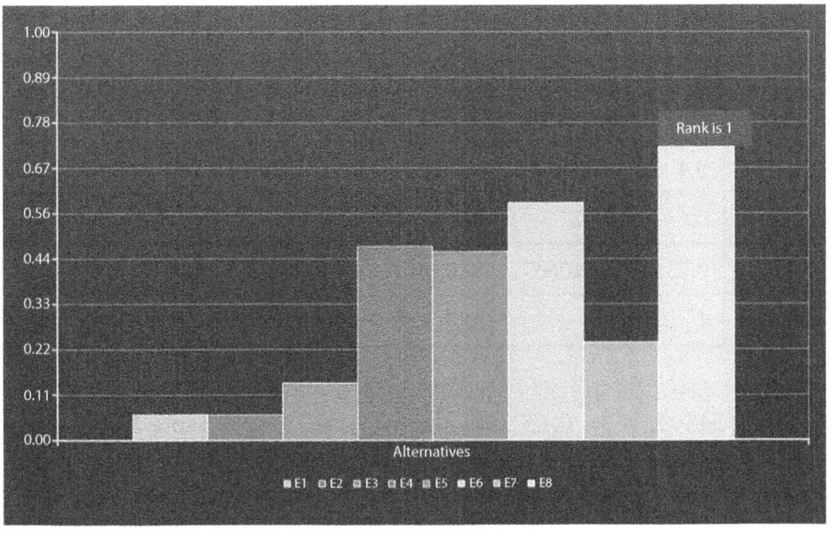

Figure 15.1 Graphical presentation of the rank performance of alternatives for NTM selection with the m-polar TOPSIS and AHP weight calculation.

Table 15.9 Decision matrix, example from [5].

Sr. no.	NTM processes	TSF	MRR	PR	C	F	M
A_1	AJM	2.5	0.8	0.22	1	1	4
A_2	USM	1.0	300	2.4	2	1	4
A_3	CHM	3.0	15	0.4	3	1	4
A_4	EBM	2.5	1.6	0.2	4	4	4
A_5	LBM	2.0	0.1	1.4	3	4	4
A_6	ECM	3.0	1500	100	5	5	4
A_7	EDM	3.5	800	2.7	3	1	5
A_8	PAM	5.0	75000	50	1	5	4

Table 15.10 Normalized decision matrix with linear max N1 method.

NTM	TSF	MRR	PR	C	F	M
1	0.5	1.07E-05	0.0022	0.2	0.2	0.8
2	0.2	0.004	0.024	0.4	0.2	0.8
3	0.6	0.0002	0.004	0.6	0.2	0.8
4	0.5	2.13E-05	0.002	0.8	0.8	0.8
5	0.4	1.33E-06	0.014	0.6	0.8	0.8
6	0.6	0.02	1	1	1	0.8
7	0.7	0.010667	0.027	0.6	0.2	1
8	1	1	0.5	0.2	1	0.8

$W_C = 0.154599$, $W_F = 0.157297$, and $W_M = 0.151639$. Table 15.11, shown below is weight multiplied matrix obtained with criteria weights given above and normalised decision matrix shown in Table 15.10.

A positive ideal solution for the alternatives are shown below calculated with the help of equation 15.1,

Table 15.11 Weight multiplied matrix.

NTM processes	TSF	MRR	PR	C	F	M
A_1	0.07675	2.14e-06	0.00040194	0.0309	0.03144	0.12128
A_2	0.0307	0.0008	0.0043848	0.0618	0.03144	0.12128
A_3	0.0921	4e-05	0.0007308	0.0927	0.03144	0.12128
A_4	0.07675	4.26e-06	0.0003654	0.1236	0.12576	0.12128
A_5	0.0614	2.66e-07	0.0025578	0.0927	0.12576	0.12128
A_6	0.0921	0.004	0.1827	0.1545	0.1572	0.12128
A_7	0.10745	0.00212	0.0049329	0.0927	0.03144	0.1516
A_8	0.1535	0.2	0.09135	0.0309	0.1572	0.12128

$$P_{IS} = \{(0.1535),(0.2),(0.1827),(0.1545),(0.1572),(0.1516)\}$$

A negative ideal solution for the alternative is shown below calculated with the help of equation 15.2,

$$N_{IS} = \{(0.0307),(2.66e\text{-}07),(0.0003654),(0.0309),(0.03144),(0.12128)\}$$

Table 15.12, shown below, represents Euclidean distance of alternatives from a positive ideal solution and euclidean distance of alternatives from the negative ideal solution obtained from equations (15.3) and (15.4), respectively.

Table 15.13 below shows that the final ranking obtained from the RCC of each alternative, highest the RCC higher will be the rank of alternatives.

Figure 15.2, shown above, is the graphical representation of rank performance obtained by implementing m-polar fuzzy TOPSIS and Shannon's entropy weight calculation method for selection of NTM process.

Table 15.12 Euclidean distance of alternatives from a positive ideal solution and negative ideal solution.

Sr. no.	Euclidean distance of alternatives from P_{IS}	Euclidean distances of alternatives from N_{IS}
1	$D_e(A_1, P_{IS}) = 0.333368$	$D_e(A_1, N_{IS}) = 0.04605$
2	$D_e(A_2, P_{IS}) = 0.334492$	$D_e(A_2, N_{IS}) = 0.0311706$
3	$D_e(A_3, P_{IS}) = 0.312123$	$D_e(A_3, N_{IS}) = 0.0871168$
4	$D_e(A_4, P_{IS}) = 0.286352$	$D_e(A_4, N_{IS}) = 0.140036$
5	$D_e(A_5, P_{IS}) = 0.294382$	$D_e(A_5, N_{IS}) = 0.116888$
6	$D_e(A_6, P_{IS}) = 0.207618$	$D_e(A_6, N_{IS}) = 0.261007$
7	$D_e(A_7, P_{IS}) = 0.30416$	$D_e(A_7, N_{IS}) = 0.10322$
8	$D_e(A_8, P_{IS}) = 0.156656$	$D_e(A_8, N_{IS}) = 0.281378$

Table 15.13 The rank of alternatives based on RCC.

Sr. no.	NTM process	Distance of alternatives from positive ideal solution with Shannon's entropy weight	Distance of alternatives from negative ideal solution with Shannon's entropy weight	Rank
1	AJM	0.333368	0.04605	Rank 7: E1,0.12137
2	USM	0.334492	0.0311706	Rank 8 : E2,0.0852441
3	CHM	0.312123	0.0871168	Rank 6 : E3,0.218207
4	EBM	0.286352	0.140036	Rank 3: E4,0.328424
5	LBM	0.294382	0.116888	Rank 4: E5,0.284213
6	ECM	0.207618	0.261007	Rank 2: E6,0.556963
7	EDM	0.30416	0.10322	Rank 5: E7,0.253376
8	PAM	0.156656	0.281378	Rank 1: E8,0.642366

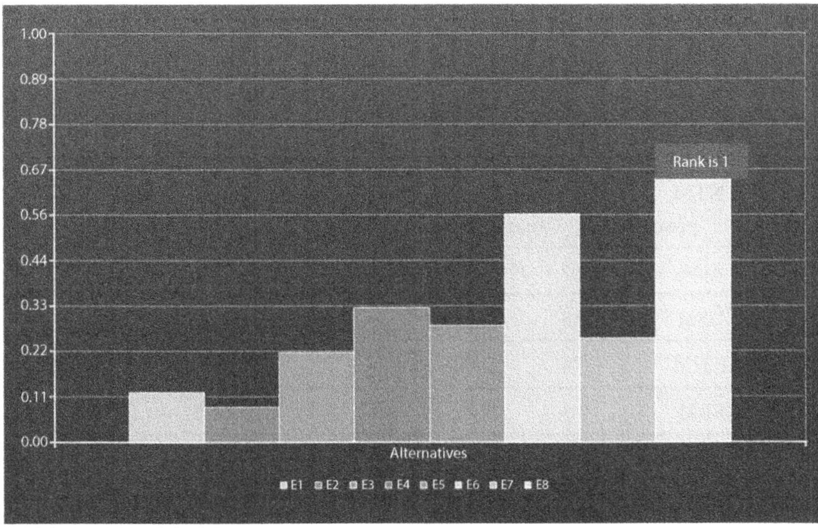

Figure 15.2 Graphical presentation of the rank performance of alternatives for NTM selection with the m-polar TOPSIS and Shannon's weight calculation method.

15.5 Results and Discussions

Applied the m-polar fuzzy TOPSIS for single pole in NTM process selection first time and obtained the rank performance for deep through cutting in Titanium, the ranking is PAM-ECM-EBM-LBM-EDM-CHM-AJM-USM.

15.5.1 Result Validation

Results are validated with the help of previous research papers. Results obtained from the m-polar fuzzy TOPSIS method are compared to the diagraph based expert system and PROMETHEE-II approach. The best alternative from all the methods is the PAM process, while the second rank is given to the ECM process in the m-polar fuzzy TOPSIS and PROMETHEE-II methods. Table 15.14 below shows the comparison of rankings for different alternatives obtained from other methodologies.

As the rank performance obtained from the m-polar fuzzy TOPSIS using AHP weight calculations and Shannon's entropy weight calculations are identical, there is a difference between positive and negative ideal solutions. Moreover, we get the difference between the distance of alternatives

Table 15.14 Result validation for m-polar fuzzy TOPSIS methodology with previous researchers work.

Sr. no.	NTM processes	NTM processes rank with m-polar fuzzy TOPSIS	NTM processes rank with diagraph based expert system	NTM processes rank with PROMETHEE-II
1	AJM	7	3	7
2	USM	8	5	4
3	CHM	6	7	6
4	EBM	3	2	5
5	LBM	4	4	8
6	ECM	2	6	2
7	EDM	5	8	3
8	PAM	1	1	1

Table 15.15 Data input uncertainty margin for a positive ideal solution with AHP and Shannon's entropy weight calculation.

Sr. no.	NTM process	Distance of alternatives from positive ideal solution with AHP weight (DPA)	Distance of alternatives from positive ideal solution with Shannon's entropy weight (DPS)	Positive uncertainty (DPS-DPA)
1	AJM	0.295271	0.333368	0.03840
2	USM	0.293306	0.334492	0.041186
3	CHM	0.287005	0.312123	0.025118
4	EBM	0.192539	0.286352	0.093813
5	LBM	0.196448	0.294382	0.097934
6	ECM	0.17203	0.207618	0.035588
7	EDM	0.276388	0.30416	0.027772
8	PAM	0.107636	0.156656	0.04902

Table 15.16 Data input uncertainty margin for a positive ideal solution with AHP and Shannon's entropy weight calculation.

Sr. no.	NTM process	Distance of alternatives from negative ideal solution with AHP weight (DNA)	Distance of alternatives from negative ideal solution with Shannon's entropy weight (DNS)	negative uncertainty (DNS-DNA)
1	AJM	0.01995	0.04605	0.0261
2	USM	0.0191853	0.0311706	0.01198
3	CHM	0.0466147	0.0871168	0.0405
4	EBM	0.174547	0.140036	−0.03451
5	LBM	0.168565	0.116888	−0.05168
6	ECM	0.238597	0.261007	0.02241
7	EDM	0.0872234	0.10322	0.01599
8	PAM	0.275586	0.281378	0.0057

from the positive ideal solution obtained from the AHP weight calculation and Shannon's entropy weight calculations. This difference is due to variation in the data input for the m-polar fuzzy TOPSIS method. It can be calculated for the distance of alternatives from negative ideal solution as well. Table 15.15 and Table 15.16 below give positive and negative uncertainty due to variation in data input.

15.6 Conclusions and Future Scope

The m-polar fuzzy TOPSIS method can be implemented to the NTM selection problem. It can be modified for single poles as shown and evaluated in case studies. AHP weight calculation and Shannon's entropy weight calculations method do not affect the rank performance of alternatives. However, the effect of criteria weight calculation on a positive and negative ideal solution affects the euclidean distance of alternatives from a positive and negative ideal solution obtained due to input data variation. The term that shows these differences are referred to here as positive uncertainty and negative uncertainty. These uncertainties further show that Shannon's entropy weight calculation is preferred over the AHP weight calculation as the relative closeness coefficient of alternatives is smaller in Shannon's entropy weight calculation compared to the AHP weight calculation method. Further, this algorithm can be implemented for various industrial problems, such as robot selection, FMS selection, and RP selection.

References

1. Chen, J., Li, S., Ma, S., Wang, X., Polar fuzzy sets an extension of bipolar fuzzy sets. *Sci. World J.*, 2014, Article ID 416530, 8, 2014.
2. Akram, M., Adeel, A., Alcantud, J.C.R., Multi-criteria group decision-making using an m-polar hesitant fuzzy TOPSIS approach. *Symmetry (Basel)*, 11, 6, 1-23, 2019.
3. Akram, M., Waseem, N., Liu, P., Novel approach in decision making with m-polar fuzzy ELECTRE-I. *Int. J. Fuzzy Syst.*, 21, 4, 1117–1129, 2019.
4. Prasad, K. and Chakraborty, S., A decision-making model for non-traditional machining processes selection. *Decis. Sci. Lett.*, 3, 4, 467–478, 2014.
5. Das Chakladar, N., Das, R., Chakraborty, S., A digraph-based expert system for non-traditional machining processes selection. *Int. J. Adv. Manuf. Technol.*, 43, 3–4, 226–237, 2009.
6. Karande, P. and Chakraborty, S., Application of PROMETHEE-GAIA method for non-traditional machining processes selection. *Manage. Sci. Lett.*, 2, 6, 2049–2060, 2012.
7. Chatterjee, P. and Chakraborty, S., Advanced manufacturing systems selection using ORESTE method. *Int. J. Adv. Oper. Manage.*, 5, 4, 337, 2013.
8. Chakladar, N.D. and Chakraborty, S., A combined TOPSIS-AHP-method-based approach for non-traditional machining processes selection. *Proc. Inst. Mech. Eng. Part B J. Eng. Manuf.*, 222, 12, 1613–1623, 2008.
9. Chakraborty, S., Applications of the MOORA method for decision making in a manufacturing environment. *Int. J. Adv. Manuf. Technol.*, 54, 9–12, 1155–1166, 2011.
10. Choudhury, T., Das, P.P., Roy, M.K., Shivakoti, I., Ray, A., Pradhan, B.B., Selection of non-traditional machining process: A distance-based approach. *IEEE Int. Conf. Ind. Eng. Eng. Manage.*, 2013, 852–856, 2014.
11. Das, S. and Chakraborty, S., Selection of non-traditional machining processes using analytic network process. *J. Manuf. Syst.*, 30, 1, 41–53, 2011.
12. Roy, M.K., Ray, A., Pradhan, B.B., Non-traditional machining process selection using integrated fuzzy AHP and QFD techniques: A customer perspective. *Prod. Manuf. Res.*, 2, 1, 530–549, 2014.

13. Sadhu, A. and Chakraborty, S., Non-traditional machining processes selection using data envelopment analysis (DEA). *Expert Syst. Appl.*, 38, 7, 8770–8781, 2011.
14. Petković, D. and Radenković, G., Selection of the most suitable non-conventional machining processes for ceramics machining by using MCDMs. *Sci. Sinter.*, 47, 2, 229–235, 2015.
15. Boral, S. and Chakraborty, S., A case-based reasoning approach for non-traditional machining processes selection. *Adv. Prod. Eng. Manage.*, 11, 4, 311–323, 2016.
16. Jagtap, M. and Karande, P., Rank assessment of robots using m-polar fuzzy ELECTRE-I algorithm, in: *Proceedings of the International Conference on Industrial Engineering and Operations Management*, Bangalore, India, August 16-18, 2021.
17. Jagtap, M., Karande, P., More, N., Decision making framework considering uncertainty. *Int. J. Adv. Sci. Technol.*, 29, 4, 7640–7649, 2020.
18. Jagtap, M. and Karande, P., Effect of normalization methods on rank performance in single-valued m-polar fuzzy ELECTRE-I algorithm. *Materials Today: Proceedings*, 2021.

16

Comparative Analysis on Material Handling Device Selection Using Hybrid FMCDM Methodology

Bipradas Bairagi

Department of Mechanical Engineering, Haldia Institute of Technology, Haldia, India

Abstract

Nowadays, manufacturing industry worldwide is facing ferocious competition, due to which its existence is under question mark. In this circumstance, decision makers in the industry are constantly searching the way of making right decision in every aspect. Right selection of material handling device (MHD) is one of the major area of decision making. This chapter aims to make right decision by combining a number of conventional FMCDM tools, *viz*. FAHP, FTOPSIS, FMOORA, FVIKOR, FGT, and COPRAS-G techniques through group decision-making process to select the best material handling device. FAHP is used to evaluate criteria weights. To take the merits of diverse evaluation processes, this chapter systematically and reasonably combines all the five remaining tools to introduce a novel super-hybrid approach with the help of entropy method. Solution of an empirical problem on MHD selection is illustrated which establishes the applicability of the approach. The study finds that the present approach is useful and effective to select the best material handling device in FMCDM environment.

***Keywords*:** FMCDM, fuzzy TOPSIS, fuzzy MOORA, fuzzy VIKOR, fuzzy grey theory, COPRAS-G

Email: bipradasbairagi79@gmail.com

Sachi Nandan Mohanty, Prasenjit Chatterjee and Bui Thanh Hung (eds.) Fuzzy Computing in Data Science: Applications and Challenges, (287–304) © 2023 Scrivener Publishing LLC

16.1 Introduction

Selection of proper Material Handling Device (MHD) is one of the most imperative managerial decisions. In general, decision makers judge a number of significant criteria while selecting MHD for industrial purpose. However, proper selection of MHD for industrial application is extremely essential. The followings are the brief literature survey in the broad areas of material handling equipment selection published in different open journals.

Goswami and Behera made an investigation for the capability and applicability of two well-known MCDM approaches ARAS and COPRAS for the evaluation and selection of conveyors, AGV and robots as material handling equipment [1]. Soufi *et al.* introduced an AHP based MCDM methodology for the evaluation and selection of material handling equipment to be utilized in manufacturing systems [2]. Satyam *et al.* applied a multi attribute decision making approach for evaluation and selection of conveyors as material handling equipment [3]. Nguyen *et al.* advocated a combined multi criteria decision making model for the evaluation and selection of conveyor as the material handling equipment based on fuzzy analytical hierarchy process and fuzzy ARAS with vague and imprecision information [4]. Mathewa and Sahua made a comparison among the novel multicriteria decision making approaches by solving a problem on material handling equipment selection [5].

Poon *et al.* used genetic algorithm to address MHE selection and operation allocation on selection of industrial conveyor [6]. Fonseca *et al.* proposed a knowledge-based method for selection of conveyor equipment as material handling equipment [7]. Tuzkaya *et al.* integrated FMCDM methodologies such as ANP and PROMETHEE utilizing fuzzy set for MHE selection [8]. Chatterjee *et al.* solved and compared robot selection problem using two MCDM techniques [9]. Shih developed TOPSIS-based group decision algorithm to choose robots applying incremental based benefit–cost ratio [10].

Kahraman *et al.* proposed a fuzzy hierarchical TOPSIS model for selection of industrial robotic systems [11]. Sujono and Lashkari proposed a method for simultaneously determining the operation allocation and material handling system [12]. Kulak developed a decision support system for MHE selection under fuzzy environment FUMAHES [13]. Lashkari *et al.* suggested an MH selection and operation allocation model in a cellular based manufacturing system [14].

Chan *et al.* developed a smart MHE selection advisor (MHESA) [15]. Chu and Lin proposed a FTOPSIS approach where the objective criteria values were transformed into index without dimension to make sure compatibility among linguistic terms and crisp value of subjective and objective criteria respectively. Through internal arithmetic of fuzzy numbers, the defuzzying of weighted rating into crisp values by the ranking method of mean of removals and determination of closeness coefficient, the robots were ranked [16].

The above literature survey reveals that some of the past researchers have applied MCDM techniques for solving material handling device (MHD) selection problem. However, this endeavor is inadequate for exhaustive and extensive decision making regarding proper selection of MHD from thousands of available alternatives under multiple conflicting criteria of diverse categories. Therefore an effort in the chapter is made to suggest appropriate techniques in making right decision by combining a number of conventional FMCDM tools, viz. FAHP, FTOPSIS, FMOORA, FVIKOR, FGT, and COPRAS-G techniques through group decision-making process to select the best material handling devices.

This research has the following objectives: (a) Comparative study of the results of the MHD evaluation problem (b) Aiding decision makers by providing the way of making more accurate decision.

The rest of the chapter is divided into several subsections. Section 16.2 describes FAHP provide Entropy methods as weight evaluation techniques. Section 16.3 briefly states the steps of six MCDM methodologies. A practical problem on material handling devices selection is specified in section 16.4. The next section 16.5 illustrates the calculation along with discussion with some salient points of the chapter. And last, Section 16.6 is dedicated to make suggestions for the direction of future research.

16.2 MCDM Techniques

16.2.1 FAHP

The concept of FAHP is briefly stated below.

Step 1: Frame a weight matrix in linguistic terms assessed by comparison between every pair of criteria.
Step 2: Transform the linguistic terms into respective fuzzy numbers.
Step 3: Build fuzzy average weight matrix.

Step 4: Calculate Geometric Mean (GM) of the fuzzy weights for each criterion.
Step 5: Calculate fuzzy weights for each criterion.

16.2.2 Entropy Method as Weights (Influence) Evaluation Technique

Entropy weighting method has been utilized to assess the objective (quantitative) weights of the FMCDM tools those are fused together to construct the proposed super hybrid approach. The step by step process of entropy method is illustrated below [17].

Step 1. Carry out the normalization of scores/ratings of the decision matrix using the following equation.

$$z_{ij} = \frac{x_{ij}}{\sum_{i=1}^{m} x_{ij}},$$

z_{ij} is normalized crisp value, x_{ij} crisp rating of ith alternative by jth method. Here i, j stand for a natural number less than or equal to m and n respectively, m is alternative number and n is criteria number.
Step 2. Calculate the entropy of each criterion with the following equation.

$$e_j = -\frac{1}{\ln m}\sum_{i=1}^{m} z_{ij} \ln z_{ij}$$

Step 3. Calculate weight (importance) of the jth criterion by the equation.

$$w_j = \frac{1-e_j}{\sum_{j=1}^{n}(1-e_j)}$$

w_j stands for objective (crisp) weight for jth criterion (here, methods to be integrated) and $\sum_{j=1}^{n} w_j = 1$.

16.3 The Proposed Hybrid and Super Hybrid FMCDM Approaches

The conception of fuzzy set theory when applied in MCDM environment is termed FMCDM [18]. FMCDM approach can support in decision making by evaluating, selecting the most excellent alternative. This chapter integrates five FMCDM techniques (Fuzzy TOPSIS, Fuzzy MOORA, Fuzzy VIKOR, Fuzzy Grey Theory and COPRAS-G) independently with FAHP for the purpose.

Five integrated techniques have been combined together with entropy method to construct the super hybrid algorithm which is constructed to take the merits of diverse evaluation procedures. The abovementioned five well-known MCDM techniques are used in MHD selection due to their wide and successful applications by past researchers in finding the best alternatives. Hybrid FMCDM approaches are described in subsection 16.3.1 to 16.3.5 whereas the proposed super hybrid approach is furnished in subsection 16.3.6.

16.3.1 TOPSIS

TOPSIS was developed and advocated by Hwang and Yoon [19]. FTOPSIS is the integration of theory of fuzzy set with TOPSIS. In past several researchers extended the TOPSIS in fuzzy environment. TOPSIS is a reference point approach. TOPSIS/FTOPSIS is on the basis of the idea that chosen alternative be supposed to have the least Euclidian distance from PIS and the highest Euclidian distance from the NIS. In this technique a worse score of one criterion can be canceled out by a better score of another criterion. Since TOPSIS/FTOPSIS permit trade-offs between criteria hence it is a compensatory method. TOPSIS/FTOPSIS gives an additional sensitive form of representation which takes in or rule out alternative resolution based on rigid cut-offs. The practical steps of FTOPSIS technique are stated below.

Step1: Build the decision matrix consisting of fuzzy numbers.
Step 2: Make the weight matrix consisting of fuzzy numbers.
Step 3: Perform the normalization process the fuzzy ratings.
Step 4: Compute the weighted normalized decision matrix with fuzzy numbers.
Step 5: Find the fuzzy PIS and Fuzzy NIS.
Step 6: Determine separation measures, positive and negative, for each alternative.

Step 7: Find relative closeness for each alternative.
Step 8: Determine the modified closeness coefficients by using the normalizing equation.

$$MCC_i = \frac{CC_i - CC_{\min}}{CC_{\max} - CC_{\min}}$$

Where, MCC_i is the modified closeness coefficient of alternative i, CC_i is the closeness coefficient of alternative i, $CC_{\max}$ is the maximum modified closeness coefficient of the alternatives and $CC_{\min}$ is the minimum closeness coefficient of the alternatives. The value of MCC_i ranges from 0 to 1; in other words, $0 \leq MCC_i \leq 1$.
Step 9: Select the best alternative with the highest MCC_i and identify the worst alternative with the least MCC_i.

16.3.2 FMOORA Method

Brauers and Zavadakas was first to introduce MOORA method [20]. **FMOORA** or fuzzy MOORA method is the concept of integration of Fuzzy set with classical MOORA. FMOORA is composed of of the following steps.

Step 1: Construction of decision matrix
Step 2: Normalization of the fuzzy performance ratings
Step 3: Determination of net score
Step 4: Best non-fuzzy performance (BNP) of each alternative is computed
Step 5: The best alternative has the maximum BNP value and the worst alternative has the minimum BNP value.

16.3.3 FVIKOR

VIKOR (Visekriterijumsko KOmpromisno Rangiranje) is an MCDM method associated with optimization and compromise. VIKOR was introduced by Zeleny [21]. VIKOR method is stated below.

Step 1: Decision matrix is constructed with fuzzy performance rating.
Step 2: PIS and NIS are determined of each criterion.

Step 3: Determine the optimal solution of alternatives' comprehensive evaluation (S_i) and the worst solution of alternatives' comprehensive evaluation (R_i).
Step 4: Compute the value of Q_i for each alternative.
Step 5: Arrange the alternatives as per the increasing sequence of defuzzyfied S_i, R_i and Q_i values. The alternative with the least value corresponds to the best one and alternative with the largest value is associated with the highest rank. Thus, three ranks of the alternatives are obtained.
Step 6: Check whether both of the state of acceptable advantage and stability in decision-making simultaneously satisfy.

16.3.4 Fuzzy Grey Theory (FGT)

Fuzzy grey theory is an efficient technique for analysis of mathematical systems characterized by uncertain information and insufficient data in fuzzy environment [22]. The procedural steps with various normalization techniques used for different sense of the criteria are furnished below [23].

Step 1: Reference data is set. Reference data consists of the most favorable value or target value for each criterion.
Step 2: Performance values for each alternative with respect to each criterion determined.
Step 3: Normalization of performance values is accomplished.
Step 4: The differences between data sets are computed. Global maximum and global minimum for each criterion are determined.
Step 5: Grey relation coefficient is evaluated.
Step 6: The fuzzy grey relational grade (GRG) is computed.
Step 7: The alternative having the maximum value of objective GRG is the most excellent alternative and the alternative with the minimum objective GRG is the worst alternative.

16.3.5 COPRAS -G

The concept of integration of Gray system theory with COPRAS (Complex Proportional assessment) is termed as COPRAS-G. COPRAS-G is a relatively fresh technique for the alternative appraisal in interval considering multiple criteria. COPRAS-G method has the following simple steps.

Step1: Form a decision matrix consisting of performance rating of alternatives in interval.
Step 2: Carryout the normalization of performance rating.
Step 3: Construct the weight matrix.
Step 4: Compute weighted normalized decision matrix.
Step 5: Calculate the average value of weighted normalized rating for every alternative with respect to each benefit criterion.
Step 6: Compute the average value of weighted normalized rating for every alternative with respect to each cost criterion.
Step 7: Evaluate relative weight of every alternative.
Step 8: Determine the degree of utility for every alternative.
Step 9: Determine modified utility index by using the normalizing equation.
Step 10: Select the alternative having the highest value of utility index or modified utility index.

16.3.6 Super Hybrid Algorithm

In this research, a super hybrid algorithm is introduced by integrating FTOPSIS, FMOORA, FVIKOR, fuzzy grey theory, and COPRAS-G. Each of these MCDM tools has computational and functional capability for finding the best alternative. To incorporate all these merits into the selection index of the paramount alternative, the research advocates the use of the following formula.

$$I_i = w_{FT} \otimes BNP_{FT} + w_{FM} \otimes BNP_{FM} + w_{FV} \otimes BNP_{FV}^{-1} + w_{FG} \otimes BNP_{FG} + w_{CG} \otimes BNP_{CG}$$

where, the notations represent the following implication.

I_i denotes the objective measure of ith MHD selection index. The higher the value of I_i, the better is the MHD. Hence higher value of I_i is desirable. The alternative associated with the highest value is the best solution. Where, w_{FT}, w_{FM}, w_{FV}, w_{FG} and w_{CG} denote the weight of fuzzy MOORA, fuzzy VIKOR, fuzzy grey theory, and COPRAS-G respectively. Entropy is considered as weight measuring method due to its inherent ability to calculate weights based on decision data, i.e. BNP obtained by the tools.

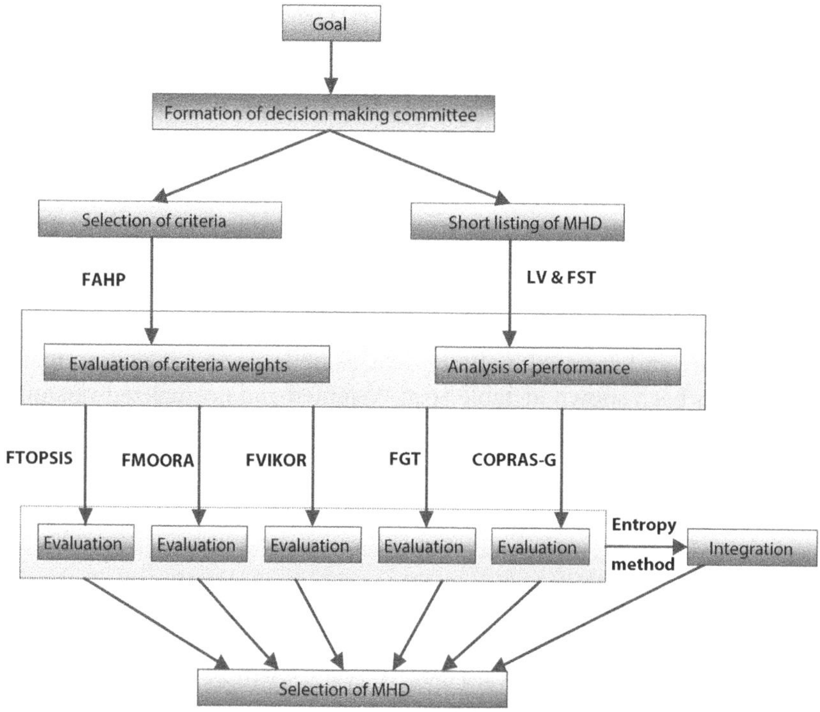

Figure 16.1 Decision making framework for MHD selection.

A careful observation shows that the weights are directly proportional to dispersion of the BNP values. BNP_{FT}, BNP_{FM}, BNP_{FV}, BNP_{FG} and BNP_{CG} are the BNP of ith MHD, obtained by respective methods. In the above equation, BNP_{FV}^{-1}(the reciprocal of BNP_{FV}) is used as it is associated with benefit value.

An evaluation framework for the selection of MHD is shown in Figure 16.1. These six proposed methodologies are illustrated with a real life problem of MHD selection under FMCDM environment. The problem on MHD selection is defined in the following section.

16.4 Illustrative Example

An Eastern Indian manufacturing organization needs to decide on the top MHD before ordering it to the vendors. The organization makes a team/committee which preliminary screens five different material handling devices. The committee unanimously considers six selection criteria.

The five MHDs are denoted by MHD1, MHD2, MHD3, MHD4 and MHD5. Six selection criteria under consideration are C1: Cost, C2: Floor/Aisle space requirement, C3: Limitation/constraints, C4: Load Capacity, C5: Traveling Distance and C6: Life Period.

C1, C2, C3 are criteria of cost category. C4, C5 and C6 are criteria of benefit category. Subjective criteria are assessed by linguistic terms which are transformed into fuzzy numbers as per Table 16.1. Handbook of MH system design is the source of assess data. Fuzzy numbers, linguistic variables, TFN, $(TFN)^{-1}$ for criteria weights are accommodated in Table 16.2. Cost is objective criterion, and expressed in crisp number. Remaining five criteria are subjective in nature and assessed by fuzzy data, as in Table 16.3. Decision matrix in TFNs has been shown in Table 16.4. Weighted and normalized version of

Table 16.1 Fuzzy numbers, linguistic variables, and scale of TFNs for ratings.

Fuzzy numbers	Linguistic variables	Scale of TFNs
$\tilde{1}$	Very poor	(0, 1, 3)
$\tilde{3}$	Poor	(1, 3, 5)
$\tilde{5}$	Medium	(3, 5, 7)
$\tilde{7}$	Good	(5, 7, 9)
$\tilde{9}$	Very good	(7, 9, 10)

Table 16.2 Fuzzy numbers, linguistic variables, TFN, $(TFN)^{-1}$ for criteria weights.

Fuzzy numbers	Linguistic variables	Scales of TFN	$(TFN)^{-1}$
$\tilde{1}$	Equally Important	(1, 1, 2)	(1/2, 1, 1)
$\tilde{3}$	Weakly Important	(2, 3, 4)	(1/4, 1/3, 1/2)
$\tilde{5}$	Essentially Important	(4, 5, 6)	(1/6, 1/5, 1/4)
$\tilde{7}$	Very Important	(6, 7, 8)	(1/8, 1/7, 1/6)
$\tilde{9}$	Absolutely Important	(8, 9, 9)	(1/9, 1/9, 1/8)

Table 16.3 Decision matrix for MHD selection problem.

B/C	(C)	(C)	(C)	(B)	(B)	(B)
Criteria →	Cost ($)	Floor space	Limitation	Load capacity	Traveling distance	Life period
MHD1	450	Good	Good	Very Good	Poor	Medium
MHD2	79.5	Poor	Medium	Very low	Poor	Good
MHD3	520	Poor	Very Good	Poor	Medium	Medium
MHD4	377	Poor	Medium	Medium	Poor	Medium
MHD5	502	Medium	Poor	Good	Medium	Medium

Table 16.4 Decision matrix in terms of triangular fuzzy numbers.

MHDi	[3]Cost	Floor space	Limitation	Load capacity	Travelling distance	Life period
MHD1	(405,450,495)	(5,7,9)	(5,7,9)	(7,9,10)	(1,3,5)	(3,5,7)
MHD2	(71.5,79.5,87.4)	(1,3,5)	(3,5,7)	(0,1,3)	(1,3,5)	(5,7,9)
MHD3	(468,520,572)	(1,3,5)	(7,9,10)	(1,3,5)	(3,5,7)	(3,5,7)
MHD4	(339,377,415)	(1,3,5)	(3,5,7)	(3,5,7)	(1,3,5)	(3,5,7)
MHD5	(451,502,552)	(3,5,7)	(1,3,5)	(5,7,9)	(3,5,7)	(3,5,7)

[3]Cost data are transformed into TFNs by 10% expansion in both sides.

Table 16.5 Weighted normalized performance ratings.

MHD	Cost	Floor space requirement	Limitations	Load capacity	Traveling distance	Life period
MHD1	(0.30, .49, 0.72)	(0.24, .48, 0.78)	(0.04, 0.18, 0.42)	(0.44, 0.75, 1)	(0.06, .24, 0.49)	(0.08, .24, .48)
MHD2	(0.05, .09, 0.13)	(0.05, .20, 0.44)	(0.02, 0.13, 0.33)	(0, 0.08, 0.3)	(0.06, .24, .49)	(0.13, .33, 0.61)
MHD3	(0.34, .53, 0.84)	(0.05, .20, 0.44)	(0.06, 0.23, 0.47)	(0.06, 0.25, 0.5)	(0.17, .40, .68)	(0.08, .24, .48)
MHD4	(0.29, .42, 0.61)	(0.05, .20, 0.44)	(0.02, 0.13, 0.33)	(0.19, 0.42, 0.7)	(0.06, .24, .49)	(0.08, .24, .48)
MHD5	(0.33, .55, 0.81)	(0.14, 0.34, 0.61)	(0.01, 0.09, 0.24)	(0.32, 0.58, 0.9)	(0.17, .40, .68)	(0.08, .24, 0.48)
PIS	(0, 0, 0)	(0, 0, 0)	(0, 0, 0)	(1, 1, 1)	(1, 1, 1)	(1, 1, 1)
NIS	(1, 1, 1)	(1, 1, 1)	(1, 1, 1)	(0, 0, 0)	(0, 0, 0)	(0, 0, 0)

decision matrix is in Table 16.5. The problem on MHD selection is solved by the proposed method for the best solution. A systematic study of the result along with short discussions is described in the subsequent section.

16.5 Results and Discussions

FAHP is employed to compute criteria weights in fuzzy form. Entropy method is utilized to calculate the influence of five FMCDM tools to construct the super hybrid approach. The FMCDM techniques along with their super hybrid approach evaluate and select the most excellent alternative. A very short description of the solution by the suggested tools has been given from subsection 16.5.1 to 16.5.6.

16.5.1 FTOPSIS

In FTOPSIS methodology, normalization operation of fuzzy assessment of devices has been accomplished. Weighted normalized rating, PIS, NIS, Separation measures and closeness coefficients are determined. The ranking of the devices by FTOPSIS is MHD2>>MHD5>>MHD4>>MHD1>>MHD3. In FTOPSIS, MHD 2 attains the maximum closeness coefficient (0.5438) which ensures maximum proximity to PIS. Therefore, MHD 2 is the best material handling device as per FTOPSIS.

16.5.2 FMOORA

Decision matrix in weighted normalized version in MOORA is identical as found in TOPSIS. Sum of weighted normalized values under benefit and non benefit criteria, and net scores are evaluated. The ranking of the devices is MHD2>>MHD5>>MHD4>>MHD1>>MHD3. In MOORA, MHD 2 has the highest net score (0.80) that ensures its maximum net benefit. Hence, with MOORA method, MHD 2 is the material handling device.

16.5.3 FVIKOR

FVIKOR ranks the MHDs based on S_i, R_i, and Q_i values. MHDs ranking order cas per S_i values is MHD2>>MHD5>>MHD3>>MHD4>>MHD1. As per R_i values the rank is MHD4>>MHD3>>MHD5=MHD1>>MHD2. As per Q_i values the ranking of the MHDs is MHD2>>MHD5>> MHD3>>MHD4>>MHD1. In FVIKOR, MHD 2 has the maximum S_i, R_i, and Q_i values which satisfy both the conditions of acceptable advantage and acceptable stability ensuring maximum proximity to the optimum solution. Thus, as per FVIKOR method, MHD 2 is regarded the best material handling device.

16.5.4 Fuzzy Grey Theory (FGT)

In FGT analysis, fuzzy rating is normalized. The deviation between data sets is calculated. These deviations, global maximum and minimum are determined. Every data point is changed to grey relational coefficient. Grey relational grade for every MHD is calculated. It is evident that the ranking of the NIPVs is MHD2>>MHD4>>MHD5>>MHD1>>MHD3. Hence, by fuzzy grey theory analysis, MHD 2 with the highest gray relational grade is the best material handling device.

16.5.5 COPRAS-G

In COPRAS-G method, fuzzy ratings are normalized. Weighted normalized rating is also calculated. Average values under benefit and cost criteria are computed. Relative weight of every MHD is calculated. Utility degree is computed. In COPRAS-G, the ranking order of the MHDs is MHD2>>MHD4>>MHD5>>MHD3>>MHD1. So, as per solution by COPRAS-G, MHD 2 with the maximum utility degree is considered as the best one. It is evident from the solutions by five MCDM techniques (Fuzzy TOPSIS, Fuzzy MOORA, Fuzzy VIKOR, Fuzzy Grey Theory and COPRAS-G) that MHD 2 is the best non-power industrial vehicle. But other MHDs have inconsistent ranking order.

16.5.6 Super Hybrid Approach (SHA)

The proposed super hybrid approach assimilates five integrated FMCDM techniques. The decision matrix for this methodology comprises of the BNPs previously obtained by the five FMCDM approaches. This method presumes unequal influences (weights) of the five FMCDM methods. The entropy method is utilized to calculate the individual influence in terms of objective weight by employing the BNPs matrix. The weights of the FMCDM tools by entropy method are computed as, $w_{FT} = 0.0019$, $w_{FM} = 0.8136$, $w_{FV} = 0.0818$, $w_{FG} = 0.0113$, and $w_{CG} = 0.0914$ respectively. The BNP values and the influences of the FMCDM methods are integrated to compute MHD selection indices (I_i) and are shown in Table 16.6. The MHD selection index (I_1) for MHD1 can be calculated as shown below.

Table 16.6 MHD selection indices and ranking by proposed super hybrid approach.

	FTOPSIS	FMOORA	FVIKOR	FGT	COPRAS-G	I_i	Ranking
Weights	0.0019	0.8136	0.0818	0.0113	0.0914	---	---
MHD1	0.5049	0.13	0.9024	1.4353	2.2162	0.3993	4
MHD2	0.5438	0.8	0.4024	1.8641	4.8893	1.1527	1
MHD3	0.486	-0.3	0.6737	1.4313	2.6805	0.0731	5
MHD4	0.5212	0.41	0.7486	1.5576	3.4131	0253	3
MHD5	0.5363	0.73	0.5367	1.5335	2.7631	0.9087	2

$$I_1 = 0.0019 \times 0.5049 + 0.8136 \times 0.1300 + 0.0818 \times 0.9024 + 0.0113 \times 1.4353 + 0.0914 \times 2.2162 = 0.3993$$

Similarly, the other MHD selection indices are $I_2 = 1.1527$, $I_3 = 0.0731$, $I_4 = 0.7253$ and $I_5 = 0.9087$.

As seen from the above calculation, the MHD2 has the highest MHD selection index (I_2 = 1.1527) MHD3 has the least value of selection index (I_3 = 0.0731). The ranking of the MHDs obtained by the SHA is 5-1-4-2-3. Therefore, the super hybrid approach clearly indicates that MHD2 is the best alternative.

The analysis of the results shows that MHD2 is the best material handling device in all the methods. The ranks of the MHDs by FMOORA, FTOPSIS and the proposed super hybrid approach are exactly identical. The ranking orders of the MHDs by FVIKOR, fuzzy grey theory COPRAS-G almost matches. A number of MHD has different ranks owing to dissimilar measures. The graph represents the ranks of the MHDs by conventional is shown in Figure 16.2. Ranking order by super hybrid approach is shown in Figure 16.3. The average value of Spearman rank correlation coefficients found to be 0.85, which is well within the acceptable limit (Table 16.7).

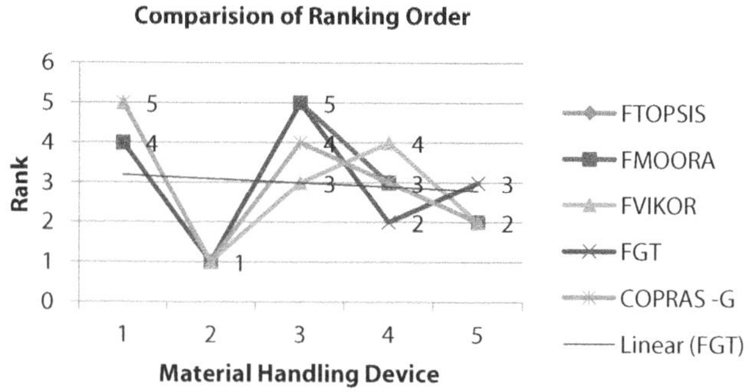

Figure 16.2 Comparison of ranking order.

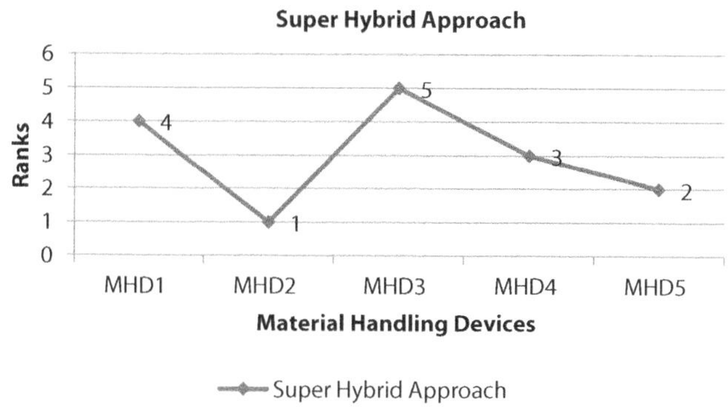

Figure 16.3 Ranks of MHDs by super hybrid approach.

Table 16.7 Spearman rank correlation coefficients.

Methodologies	FTOPSIS	FMOORA	FVIKOR	FGT	COPRAS-G	SHA	Average value
FTOPSIS	1	1	0.7	0.9	0.8	1	0.85
FMOORA	1	1	0.7	0.9	0.8	1	
FVIKOR	0.7	0.7	1	0.5	0.7	0.7	
FGT	0.9	0.9	0.5	1	0.9	0.9	
COPRAS-G	0.8	0.8	0.7	0.9	1	0.8	
SHA	1	1	0.7	0.9	0.8	1	

16.6 Conclusions

Selection of material handling devices (MHD) is a significant function of decision makers in the manufacturing industry. In many circumstances, the success and existence of manufacturing unit/organization rely on the proper selection of material handling equipment, especially nonpowered industrial vehicles. Hence, evaluation and selection of non-powered industrial vehicles is essential though adequate investigation regarding this subject has not been reported in the open journals. The current work attempts to mitigate the literature gaps. For the purpose, a decision problem on available MHDs is constructed in the unconstructed, uncertain, ambiguous and fuzzy environment. In such condition, theory of fuzzy set is a suitable means to cope with this sort of problems. In actual decision-making procedure, the decision maker is incapable/reluctant to state his/her assessment accurately in numerical value, and the estimation is expressed in linguistic terms.

The application of five FMCDM techniques (Fuzzy TOPSIS, Fuzzy MOORA, Fuzzy VIKOR, Fuzzy Grey Theory and COPRAS-G) along with super hybrid approach reveals the best MHD. The consequence of the evaluation process found by various optimization approaches shows the conformity in finding the best solution. The proposed six methodologies illustrated in the present paper can be utilized in the future managerial decision making procedures in the practical case. The result ensures that the FMCDM methodologies including super hybrid approach as useful and effective FMCDM tools in evaluation and selection of MHDs. Utilization of different conversion scales, different fuzzy numbers, and incorporation of fuzzy performance values in the super hybrid algorithm might be directions of further research.

References

1. Goswami, S.S. and Behera, D.K., Solving material handling equipment selection problems in an industry with the help of entropy integrated COPRAS and ARAS MCDM techniques, in: *Process Integration Optimization and Sustainability*, vol. 5, pp. 947–973, 2021.
2. Soufi, Z., David, P., Yahouni, Z., A methodology for the selection of material handling equipment in manufacturing systems, in: *IFAC-Papers Online*, vol. 54, pp. 122–127, 2021.

3. Satyam, F., Satywan, K., Avinash, K., Application of multi-attribute decision-making methods for the selection of conveyor. *Res. Square,* 1–18, 2021.
4. Nguyen, H.-T.N., Siti, D., Nukman, Y., Hideki, A., An integrated MCDM model for conveyor equipment evaluation and selection in an FMC based on a fuzzy AHP and fuzzy ARAS in the presence of vagueness. *PloS One*, 11, 2016.
5. Mathewa, M. and Sahua, S., Comparison of new multi-criteria decision making methods for material handling equipment selection. *Manage. Sci. Lett.*, 8, 139–150, 2018.
6. Poon, T.C., Choy, K.L., Cheng, C.K., Lao, S.I., Lam, H.Y., Effective selection and allocation of material handling equipment for stocha production material demand problems using genetic algorithm. *Exp. Syst. Appl.*, 38, 12497–12505, 2011.
7. Fonseca, D.J., Uppal, G., Greene, T.J., A knowledge-based system for conveyor equipment selection. *Exp. Syst. Appl.*, 26, 615–623, 2004.
8. Tuzkaya, G., Gülsün, B., Kahraman, C., Özgen, D., An integrated fuzzy multi-criteria decision making methodology for material handling equipment selection problem and an application. *Exp. Syst. Appl.*, 37, 2853–2863, 2010.
9. Chatterjee, P., Athawale, V.M., Chakraborty, S., Selection of industrial robots using compromise ranking and outranking methods. *Robot. Comput. Integr. Manuf.*, 26, 5, 483–489, 2010.
10. Shih, H.S., Incremental analysis for MCDM with an application to group TOPSIS. *Eur. J. Oper. Res.*, 186, 720–734, 2008.
11. Kahraman, C., Cevik, S., Ates, N.Y., Gulbay, M., Fuzzy Multi-criteria evaluation of industrial robotic systems. *Comput. Ind. Eng.*, 52, 414–433, 2007.
12. Sujono, S. and Lashkari, R.S., A multi-objective model of operation allocation and material handling system selection in FMS design. *Int. J. Prod. Econ.*, 105, 116–133, 2007.
13. Kulak, O., A decision support system for fuzzy multi-attribute selection of material handling equipments. *Exp. Syst. Appl.*, 29, 310–319, 2005.
14. Lashkari, R.S., Boparai, R., Paulo, J., Towards an integrated model of operation allocation and material handling selection in cellular manufacturing systems. *Int. J. Prod. Econ.*, 87, 115–139, 2004.
15. Chan, F.T.S., Ip, R.W.L., Lau, H., Integration of expert system with AHP for the design of material handling equipment system selection. *J. Mater. Process. Technol.*, 116, 137–145, 2001.
16. Chu, T.C. and Lin, Y.C., A fuzzy TOPSIS method for robot selection. *Int. J. Adv. Manufac. Technol.*, 21, 284–290, 2003.
17. Ding, J.F. and Liang, G.S., Using fuzzy MCDM to select partners of strategic alliances for liner shipping. *Inf. Sci.*, 173, 197–225, 2005.

18. Bellman, R.E. and Zadeh, L.A., Decision-making in a fuzzy environment. *Manage. Sci.*, 17, B141–B164, 1970.
19. Hwang, C.L. and Yoon, K., *Multiple attribute decision making methods and applications*, Springer-Verlag, New York, 1981.
20. Brauers, W.K.M. and Zavadakas, E.K., The MOORA method and its application to privatization in a transition economy. *Control Cybern.*, 35, 445–469, 2006.
21. Zeleny, M., *Multiple criteria decision making*, McGraw-Hill, New York, 1982.
22. Ozcan, T., Celebi, N., Esnaf, S., Comparative analysis of multi-criteria decision-making methodologies and implementation of a warehouse location selection problem. *Exp. Syst. Appl.*, 38, 9773–9779, 2011.
23. Yuan, X., Grey relational evaluation of final situation of listed company. *J. Modern Account. Audit.*, 3, 41–44, 2007.
24. Apple, J.M., *Material handling systems design*, John Willy and Sons, New York, 1972.

17

Fuzzy MCDM on CCPM for Decision Making: A Case Study

Bimal K. Jena[1], Biswajit Das[1], Amarendra Baral[2] and Sushanta Tripathy[1]*

[1]KIIT Deemed to be University, Bhubaneswar, Odisha, India
[2]Trident Academy of Technology, Bhubaneswar, Odisha, India

Abstract

Theoretical analysis of critical chain project management (CCPM) is presented and practical use of CCPM is illustrated in this study. For improving the Project Management Body of Knowledge, CCPM helps provide considerable steps. Following the CCPM approach, the improved impact is studied on the ground of schedule recording, cost, and scope performance. The critical chain aspect is faced by project management where resource dependencies and never-changing criteria are shown. CCPM assures feasibility and tolerable state at the time of statistical fluctuations. In this paper, a case study of an agricultural firm has taken near the TALADANDA canal and National Rice Research Institute (NRRI), Cuttack, Odisha, India. The practical implementation of CCPM on the agriculture firm is presented with a selection of best cropping planning multi-criteria context. The three objectives, maximization of net benefit, agricultural product and labor utilization are emphasized. Under CCPM, the decision-maker selects a few out of the whole lot by their preference and choice. Then by cluster analysis, we have reduced a large number of solutions to manageable numbers of the group each consisting of relatively homogeneous alternatives. Then both crisp and fuzzy method has applied for the final selection of the policy under consideration. Finally, fuzzy Technique for Order of Preference by Similarity to Ideal Solution (TOPSIS) is implemented for the final selection of policies for maximization of net benefit and to solve the critical chain project management.

***Keywords*:** CCPM, project schedule, cluster analysis, crisp MCDM, fuzzy TOPSIS

**Corresponding author*: sushant.tripathy@gmail.com

Sachi Nandan Mohanty, Prasenjit Chatterjee and Bui Thanh Hung (eds.) Fuzzy Computing in Data Science: Applications and Challenges, (305–328) © 2023 Scrivener Publishing LLC

17.1 Introduction

In a management system, decision making is an important and influencing job. Decision making is required in planning, organizing, and controlling. Commonly, decision-making is interpreted as an art, a talent habituated through experiences. Now, decision making is a challenging job, it is complex and fast-changing. So, systematic and scientific approaches are needed for the smooth management of the project. The following four aspects are considered for decision making flexible and appropriate.

1. Clear representation of the problem statement.
2. Collection of relevant facts.
3. Analyzation of the relevant facts properly.
4. Finding and implementing the solutions.

The multicriteria decision-making (MCDM) model is used for choosing the best from several alternatives. The alternatives have challenging potential with several concrete or vague criteria. Aruldoss *et al.* [2] has surveyed multicriteria decision-making method. The qualitative & quantitative factors are considered and evaluated by the decision maker (DM). When alternatives are known, MCDM is not associated with the problem. There are design constraints and different alternatives. The quantified objectives are acceptable by DM. The best alternative is to be found out with the thrust of the MCDM design. For CCPM, one solution is required, and we have proposed an MCDM approach. MCMD approach optimizes the evaluation of supply as well as different crops selection processes. The MCMD approach gives a sustainable and feasible consideration. Literature reviews have given ideas on criteria to follow in selecting a location. Besides, DM also determines the criteria for the selection of the location. Criteria's weight is assigned using the fuzzy TOPSIS method. TOPSIS is a MCMD and TOPSIS is used to rank the suppliers according to their potential. Finally, we have the suppliers with their respective potential ranks.

The remainder of this chapter is structured as follows: Section 17.2 presents the conceptual background of the research, based on a review of the previous studies on multi criteria decision-making methods applied to decision makers (DM) selection problems. Section 17.3 describes the objective of research. Sections 17.4 and 17.5 present different cluster methods and the clustering of the present problem. Section 17.6 describes the methodology of the present study. Sections 17.7 and 17.8 present a numerical case to verify the proposed model and the research findings by fuzzy TOPSIS method. Finally, Section 17.9 summarizes the conclusions and exposes work limitations and directions for further research.

17.2 Literature Review

According to Eliyahu M. Goldratt [4], CCPM approach is used for planning and managing the projects. This approach emphasizes the resources, like people, equipment, physical space, and these resources are the primary requirements to execute the project work. For business development, the process of selecting appropriate suppliers is a crucial task. In the globalized competition of business, sustainable business is a challenging task. To achieve such a goal, a sustainable supplier selection process has followed. The appropriate suppliers' influence in increasing the satisfaction of the customers, stand in the competitive market and optimize the cost of production. In particular, a fuzzy solution approach to supplier selection and order allocation in green supply chain considering the location-routing problem have been used by Qazvini and Haji [13].

Garcia and Lamata [11] have discussed the approach of ranking the different crops using fuzzy TOPSIS. Giusti and Marsili [3] have presented the approaches of fuzzy decision support systems for water conservation and irrigation in the field of agriculture. Hobbs [7] has shown the importance of rank coefficient correlation on implementing MCDM methods. Morse [12] showed that the clustering approaches can reduce different alternatives size into required sizes and the use of hierarchical clustering methods like Distance Between Centroids, Group Average & Ward's Method are illustrated nicely. Besides, Mainuddin *et al.* [10] have modeled an irrigation plan for the groundwater irrigation project of Thailand, where they have designed the optimal cropping pattern and groundwater abstraction requirement.

In this research work, the triangular fuzzy membership function is implemented to quantify the linguistic values of subjective preferences of the expert. Fuzzy values are generated for finding criteria weights and evaluating qualitative performance. Then fuzzy TOPSIS is applied to rank the different crops. Considering the rabi season and using the linear programming technique, an optimal plan model is proposed. The purpose of the model is to allocate the available water at the main canal among the branch and distribute canals. The model emphasizes optimizing the aggregate net return.

17.3 Objective of Research

Based on the investigation and on the considerations for command area planning, the following objectives are formulated.

a. Formulation of multiobjective model for generation of alternatives policies.
b. Selection of nondominated alternatives.
c. Cluster analysis for reducing the size of nondominated alternatives to a manageable subset.
d. Ranking of policies by Fuzzy TOPSIS method.

17.4 Cluster Analysis

When the efficient nondominated solution is generated from the process from the linear programming problem, still we get a large number of solutions which becomes difficult to use MCDM methods for the selection of best among them. In such case, cluster analysis helps to reduce a large number of solutions to the manageable number of groups each containing relatively manual groping processes, such as:

1. The application of clustering is in an objective function consisting of criteria and the clustering program formulates the groups without containing the conflicting situations.
2. The Clustering approach formulates clusters in few times although there are too many criteria with each alternative.
3. At certain values, it is very difficult to differentiate between two alternatives manually where the machine can difference. Since the clustering algorithm is speed, reliable, and consistent enough to organize data, so it gives a reasonable benefit.

Cluster analysis organized the data by abstracting the underlying structure either as a grouping of individual data or as a hierarchy of groups. The clustering approach N amounts of data into k number of groups where each group has homogeneous alternatives. All data are represented in a test space as points. The distance of points within a cluster is less compared with the distance of the points of different clusters. Cluster analyses are hierarchical and partitional. A nested sequence of partitions is called hierarchical clustering. Partitional clustering refers to a single partition. Hartigan [6] and Jain and Dubes [9] stated that hierarchical clustering is a special sequence of partitional clustering.

17.4.1 Hierarchical Clustering

To make N amount of data into k clusters, at first N groups are formed. Each group of N groups contains one sample. Then N-1 groups are formed by merging one group with another appropriate group. Then N-2 groups are formed following the same approach. A similar process has continued all alternatives from one cluster. Finally, we find out that the first level contains N clusters and level N consists of one cluster. Hierarchical clustering implies two alternatives are in level k, and it is in a higher level.

The hierarchical clustering is classified as agglomerative and divisive, depending on the procedures adopted. Agglomerative classification proceeds with a singleton group and from the sequence by successively merging the groups. According to divisive procedure, at fast all alternatives are taken in a cluster and then successively split as clusters. The computational cost in agglomerative is less in comparison to divisive procedure.

In our work, the agglomerative approach is employed to get the initial partition of groups single linkage, complete linkage, average distances within clusters, the average distance between clusters and Ward's method are the available five methods in this category. All these methods differ in how the distance matrix is updated after two clusters have joined. Morese [12] stated that in Ward's method, clusters are formed to minimize the increase of sums of squares within the cluster.

17.4.2 Partitional Clustering

In the partitional clustering approach, alternative elements of a cluster are identical inside in comparison to the alternative's elements of outer clusters. Square error criterion is used to seek optional partition by interactive optimization. These approaches guarantee local but not global optimal, particularly if alternatives are close to each other (Jain and Dubes [9]). Different initial partitions can also lead to different solutions. Despite these limitations, the minimum computational requirements make this methodology more appealing.

K-means clustering algorithm is one of the best-known techniques in partitional clustering methodology. In this approach, clusters are formed with the criteria that total squares difference within the cluster is minimum i.e., the error is minimum. Weights are assigned to alternatives of k-groups. The sum of the squared differences is used as a criteria to formulate the clusters and find the group of an alternative. Clusters are built iteratively by transferring alternative elements from one cluster to another with satisfying the criteria where the criteria are the sum of the square difference is minimum. The process of transferring is terminated when all alternatives are in the appropriate cluster with satisfying criteria. For a cluster of k groups, the total square error is calculated using equation 17.1.

$$E_i = \sum e_i^2 \text{ , } i = 1, 2, \ldots\ldots, k \tag{17.1}$$

where E_i = Total square error value for all cluster group i
e_i^2 = square error value for each cluster group i

17.5 Clustering

The method of clustering is used to reduce the proposed policies to a small number of representative subsets for proper analysis. In our case study, we have taken 37 policies. The 37 proposed policies are normalized by differences between the maximum and the minimum value of each criterion to make them dimensionless and then weighted normalized data have been calculated, which is given in Table 17.1.

These weighted normalized proposed policies are grouped in to different clusters given in Table 17.2.

Then six number of clusters by using K-means clustering algorithm have selected to calculate the highest Silhousette index. The highest value of Silhousette index is obtained for k = 3. So the ideal cluster is taken as 3. After fixing the optimal number of clusters as 3, the representative policy of each cluster is determined. For this purpose, the square error values for each criterion in that group are calculated. The summation of these square error values for all criteria gave the total square error value corresponding to each proposed policy in that group. The policy that gives the minimum total square error values is chosen as the leader policy for that group. The final selection of six policies is shown in Table 17.3.

Table 17.1 Weighted normalized data.

SL	Normalized data for net benefit	Normalized data for production	Normalized data for labor	SL	Normalized data for net benefit	Normalized data for production	Normalized data for labor
1	4.075398	2.7827819	1.178825	20	3.874853	2.8420741	1.150924
2	4.098462	2.8507326	1.131583	21	3.782134	2.8791235	1.127460
3	4.093112	2.8229123	1.150925	22	3.780826	2.8050242	1.234626
4	4.077670	2.7426512	1.206726	23	3.788639	2.8050265	1.178825
5	4.076741	2.8605989	1.130151	24	3.765996	2.9902715	1.109613
6	4.068863	2.7025208	1.234626	25	3.750825	2.9532218	1.116433
7	4.058477	2.7494505	1.206725	26	3.744980	2.8605985	1.206724
8	3.970402	2.8235492	1.178826	27	3.747738	2.8605985	1.150914
9	3.954374	2.9161732	1.122080	28	3.728573	2.8235492	1.234625
10	3.913587	2.9346978	1.119389	29	3.725080	2.9532218	1.150942
11	3.885856	2.8976445	1.124770	30	3.710512	2.9717468	1.112801
12	3.870968	2.9532212	1.116433	31	3:706478	2.9161725	1.123042
13	3.789057	2.8976445	1.124770	32	3.678794	3.0087962	1.107630

(Continued)

Table 17.1 Weighted normalized data. (*Continued*)

SL	Normalized data for net benefit	Normalized data for production	Normalized data for labor	SL	Normalized data for net benefit	Normalized data for production	Normalized data for labor
14	3.759488	2.8235488	1.206726	33	3.672679	2.9717465	1.140914
15	3.844500	2.9532242	1.123025	34	3.650744	2.9902711	1.109615
16	3.845068	2.9161741	1.122080	35	3.593589	3.0273208	1.105632
17	3.823703	2.9717466	1.112804	36	3.570550	2.9532218	1.178825
18	3.803579	2.9346945	1.119389	37	3.552285	2.9717465	1.178841
19	3.797235	2.8420736	1.206725				

Table 17.2 Grouping of policies into different clusters.

{No of Cluster		Policy numbers in final partition K-means
2.	Cluster-1	P01, P02, P03, PO4, P05, P06, P07, P08
	Cluster-2	P10, P11, P12, P13, P14, P15, P16, P17, P18, P19, P20, P21, P22, P23, P24, P25, P26, P27, P28, P29, P30, P31, P32, P34, P35, P36, P37, P38
3.	Cluster-1	P09, P10, P11, P12, P13, P14, P15, P16, P17, P18, P19, P20, P21, P22, P23, P24, P25, P26, P27, P28
	Cluster-2	P28, P29, P30, P31, P32, P33, P34, P35, P36
	Cluster-3	P02, P03, P04, PO5, P06, P07, P08, P09
4.	Cluster-1	P14, P19, P20, P21, P22, P23, P26, P27, P28
	Cluster-2	P24, P25, P29, P30, P31, P32, P33, P34, P35, P36, P37
	Cluster-3	P01, P02, P03, PO4, P04, P05, P06, P07
	Cluster-4	P10, P11, P12, P13 & P15, P16, P17, P18
5.	Cluster-1	P11, P12, P13, P15, P16, P17, P18, P21, P24, P25
	Cluster-2	PO4, P07, P08
	Cluster-3	P29, P30, P31, P32, P33, P34, P35, P36, P37
	Cluster-4	P14, P19, P20, P22, P23, P26, P27, P28
	Cluster-5	P04, P06, P07, P08
6.	Cluster-1	P36, P37
	Cluster-2	P29, P30, P31, P32, P33, P34, P35
	Cluster-3	P10, P11, P12, P13, P15, P16, P17, P18, P21, P24, P25
	Cluster-4	P14, P19, P20, P22, P23, P26, P27, P28
	Cluster-5	P02, P03, P05, P08, P09
	Cluster-6	P01, PO4, P06, P07

Table 17.3 Selection of alternative.

Chosen alternative policy	Labour	Production	Benefits	Represented policies
RP1	1.1274611	2.8791225	3.7921330	P21
RP2	1.1110800	239161735	3.8450670	P16
RP3	1.1309240	2.9717455	3.6726790	P33
RP4	1.1126130	2.8802711	3.6507550	P34
RP5	1.1828250	2.7827819	4.0853870	P1
RP6	1.1609250	2.8339123	4.0932230	P3

17.6 Methodology

Initially, the three objective function, i.e., maximization of net benefits, maximization of agricultural production and maximization of labor utilization are solved separately as single objective linear programming (LP) problem to determine the maximum and minimum values that can be achieved by each objective. The same set of constraints is used in each case. The solution relating to each objective is given in Table 17.4a. Further, the values of other two objective functions corresponding to the optimal solution of one objective function are calculated and shown in

Form of a pay-off matrix in Table 17.4b. It is found from the Table 17.1 (a) that three planning objectives are conflicting with one another. There arises necessity to develop tradeoff relationships and to select best alternative cropping plan in the multiobjective irrigation planning scenario to meet the chosen levels of satisfaction as demanded in the decision making process where X_L, X_P, and X_B denote the optimal solution for maximization of labour, production and net benefit respectively.

Table 17.4 (a) Ideal crop plan from the planning model.

Sl. no.	Crops	Represented variables	Solution for maximization of		
			Labor	Production	Benefit
1.	Jute	X_1	117.74	119.98	115.26
2.	Auspaddy	X_2	201.74	110.57	124.77
3.	Aman paddy	X_3	86.37	26.88	16.88
4.	Boro paddy	X_4	0.00	0.00	0.00
5.	Groundnut	X_5	62.67	8.87	8.87
6.	Mustard	X_6	9.38	4.39	4.40
7.	Sesamum	X_7	164.56	162.82	168.82
8.	Green gram	X_8	8.78	8.88	8.78
9.	Black gram	X_9	4.38	4.28	4.41
10.	Patato	X_{10}	112.19	0.00	0.00
11.	Tomato	X_{11}	1.88	1.89	1.75
12.	Cauliflower	X_{12}	0.00	0.00	0.00
13.	Cabbage	X_{13}	85.38	61.68	61.55
14.	Radish	X_{14}	80.58	127.65	137.56
15.	Onion	X_{15}	0.86	124.58	124.58
16.	Cucumber	X_{16}	79.04	122.89	144.89
17.	Pumpkin	X_{17}	0.86	0.86	0.88
18.	Brinjal	X_{18}	0.88	0.75	0.88
19.	Total crop area	ha	1026.32	884.04	913.52

Table 17.4 (b) Pay-off matrix.

Criteria	X_L	X_P	X_B
Net benefit (in Rs)	12077060	12307830	17172690
Agricultural Production (in Kg)	73220011	8208790	7689401
Labour utilization (in Man days)	18493.5	152222.9	161731.8

17.7 TOPSIS Method

Hwang and Yoon [8] proposed the alternative method of ELECTRE method and the approach is the TOPSIS method. Ahn, B.S [1] has defined that Euclidean geometrical distance is applied in the TOPSIS approach. The geometrical approach is used to find out (a) The shortest distance from an ideal solution, (b) The longest distance from a nonideal solution or (c) not the solution.

TOPSIS considers monotonically increasing and decreasing utility features tendency. From the tendency, ideal and nonideal solutions are identified. The Euclidean distance measure is applied to evaluate the relative closeness of alternatives to an ideal solution. Then order sequence of alternatives is depicted to sketch the preferences order from the alternatives. This preference order is made by comparing the relative distances.

The TOPSIS formulates the decision matrix as defined in the table. The matrix contains m rows and n columns, where m refers to alternatives and n refers to criteria.

Criterion

Weight / Criterion	C_1	C_2	C_3	……	C_n
	W_1	W_2	W_3	……	W_n
A_1	x_{11}	x_{12}	x_{13}	……	x_{1n}
A_2	x_{21}	x_{22}	x_{23}	……	x_{2n}
A_3	X_{31}	X_{32}	X_{33}	……	X_{3n}
……	……	……	……	……	……
A_m	x_{m1}	x_{m2}	x_{m3}	……	x_{mn}

In the table, A_i is the ith alternative, C_j is the jth criterion, and x_{ij} is the performance with ith alternative and jth criterion. Using the TOPSIS method, the steps are given below.

Step 1: Construction of normalized decision matrix:
Using the ELECTRE method, the attribute dimensions are converted into nondimensional attributes. Assuming R as the normalized decision matrix and r_{ij} as the element of R, then r_{ij} is calculated using equation 17.1.

$$r_{ij} \frac{X_{ij}}{\sqrt{\sum_{i-1}^{m} x_{ij}^2}} \tag{17.1}$$

Step 2: Construction of weighted normalized decision matrix:
DM specifies the weights set as W where

$$W = (w_{1,}\ w_{2}\ldots\ldots\ldots\ldots\ldots w_{n})\ \text{and}\ \Sigma\ w_{i} = 1.$$

The normalized decision matrix, R is also taken into consideration.
The weighted normalized matrix (V) is determined as $(v_{ij)} = (r_{ij}w_{i})$.
Step 3: Determination of ideal and negative ideal solutions:
Assume, the ideal solution as A* and the negative ideal solution as A-.
A* and A- are defined as follows.

$$A^{*} = \{(m_{i}axv_{ij}\ |\ j \in J), (m_{i}inv_{ij}\ |\ j \in j')\}\text{for } i = 1, 2, \ldots\ldots, m \tag{17.2}$$

$$= (V^{*}_{1}\ V^{*}_{2}\ldots\ldots\ldots\ V^{*}_{n})$$

$$A- = \{(m_{i}invv_{ij}\ |\ j \in J), (m_{i}axv_{ij}\ |\ j \in J')\}\ \text{for } I = 1, 2, \ldots\ldots, M \tag{17.3}$$

$$= (V_{1}^{-}\ V_{2}^{-}\ldots\ldots\ V_{n}\text{-})$$

where J = {j = 1, 2,......, n} & & j' = {j = 1, 2,........, n}.

DM intension is to have maximum value in the benefit criteria. On another side, DM intention to have minimum value is cost criteria. A* is interpreted as an ideal solution. A- is interpreted as a negative - ideal solution. A* is more preferable, whereas A- is least preferable.
Step 4: Calculating the separation measure:
Using the n-dimensional Euclidean distance formula the separation distances of each alternative are evaluated from an ideal solution and negative ideal solution. Thus, the following formulas are implemented to calculate the distance. For, i =1, 2,.........., m

$$S_i^* = \sqrt{\Sigma(V_{ij} - V_j^*)^2}$$

where s_i^* = Separation measure of alternative i from the ideal solution & where s_i^- = Separation measure of alternative i from the negative ideal solution.

Step 5: Relative closeness to the ideal solution calculation:

Dwivedi *et al.* [5] have defined the relative closeness of alternative A and the ideal solution A* as given below.

$$C_i^* = \frac{S_i^-}{S_i^* + S_i^-} 0 \leq C_i^* \leq 1$$

$$C^*_i = 1, \text{ if and only if } A_i = A^* \;\&\; C^-_I = 1 \text{ if and only if } A_1 = A^-$$

Step - 6: Ranking the preference order:

C^*_I has been ordered according to preference rank. The best satisfying alternative is selected from the preference rank order. The approach of distance measure is assuring that the selected alternative has the shortest distance from the ideal solution and on the contrary has the longest distance from the negative ideal solution.

17.8 Fuzzy TOPSIS Method

As discussed in the section 17.5 and using fuzzy triangular number, we got the fuzzy decision matrix given in Table 17.5. Then the fuzzy version of TOPSIS method is to find the rank of the alternative is described below.

Step 1: Considering the fuzzy decision matrix given in Table 17.5, the normalized decision matrix is obtained as below in Table 17.6.

Step II: Considering the normalized decision matrix and the weight of different criteria, the weighted normalized decision matrix is calculated. The same is presented in Table 17.7.

Step III: The following are the positive ideal solution & negative ideal solution obtained in the calculation.

A* = {(0.029502, 0.060446, 0.094015), (0.081236, 0.137478, 0.192164), (0.185901, 0.229836, 0.276068)}

Table 17.5 Fuzzy decision matrix.

Polices	Labour			Production			Benefit		
	Lower	**Modal**	**Upper**	**Lower**	**Modal**	**Upper**	**Lower**	**Model**	**Upper**
RP1	1.018125	1.154460	1.236654	2.733875	2.879271	3.024325	3.640852	3.792214	3.943692
RP2	1.012541	1.122281	1.231124	2.815999	2.916321	3.061359	3.693812	3.845058	3.996238
RP3	1.065842	1.120924	1.216215	2.842852	2.971654	3.053581	3.511874	3.672578	3.752499
RP4	1.034215	1.128613	1.185324	2.887376	2.990548	3.072244	3.489852	3.650651	3.778564
RP5	1.083247	1.178725	1.273658	2.640789	2.782512	2.924628	3.983248	4.085541	4.215549
RP6	1.102044	1.150875	1.245721	2.680985	2.822952	2.964781	3.990852	4.093221	4.352933

Table 17.6 Normalized decision matrix.

Coital kV	Labour			Production			Benefit		
Policy/ weight	0.0697	0.1430	0.2227	0.1897	0.3260	0.4623	0.4249	0.5310	0.6371
RPI	0.397649	0.404281	0.409854	0.402554	0.406038	0.409193	0.399147	0.401008	0.401678
RP2	0.395548	0.402352	0.408071	0.408009	0.411263	0.414205	0.404950	0.406606	0.407070
RP3	0.416074	0.409109	0.403140	0.425509	0.419100	0.413163	0.384994	0.388376	0.390493
RP4	0.403847	0.397881	0.392763	0.428237	0.421713	0.415669	0.382589	0.386057	0.388259
RP5	0.423268	0.422699	0.422158	0.388847	0.392451	0.395717	0.436671	0.432020	0.426558
RP6	0.412373	0.412695	0.412911	0.394756	0.398110	0.401147	0.437516	0.432836	0.433320

Table 17.7 Weighted normalized decision matrix.

Policy	Labour			Production			Benefit		
RP1	0.027716	0.057812	0.091275	0.076364	0.132368	0.189170	0.169598	0.212935	0.255909
RP2	0.027570	0.057536	0.090877	0.077399	0.134072	0.191487	0.172063	0.215908	0.259345
RP3	0.029000	0.058503	0.089779	0.080719	0.136627	0.191005	0.163584	0.206228	0.248783
RP4	0.028148	0.056897	0.087468	0.081236	0.137478	0.192164	0.162562	0.204996	0.247360
RP5	0.029502	0.060446	0.094015	0.073764	0.127939	0.182940	0.185541	0.229403	0.271760
RP6	0.028742	0.059015	0.091955	0.074885	0.129784	0.185450	0.185901	0.229836	0.276068

Table 17.8 Separation measures.

Separation measures	Lower	Modal	Upper	Separation measures	Lower	Modal	Upper
S_1^*	0.017109	0.017852	0.020563	S_1^-	0.007502	0.009096	0.010586
S_2^*	0.014489	0.014631	0.017029	S_2^-	0.010173	0.012517	0.014720
S_3^*	0.022329	0.023703	0.027636	S_3^-	0.007174	0.008828	0.009263
S_4^*	0.023378	0.025092	0.029445	S_4^-	0.007495	0.009561	0.009834
S_5^*	0.007480	0.009549	0.010180	S_5^-	0.23060	0.024580	0.024601
S_6^*	0.006396	0.007826	0.008023	S_6^-	0.23395	0.024952	0.028838

$A^- = \{(0.02757, 0.057536, 0.090877), (0.073764, 0.127939, 0.18294), (0.162562, 0.204996, 0.247360)\}$

Step IV: The calculation of separation measures is done as given in Table 17.8.

For instance, the calculation of S_1^* is done as

$$S_1^* = ([(0.027716, 0.057812, 0.091275) - (0.029502, 0.060446, 0.094015)^2$$

$$+ [(0.076364, 0.132368, 0.189170) - (0.081236, 0.137478, 0.192164)^2$$

$$+ [(0.169598, 0.212935, 0.255909) - (0.185901, 0.229836, 0.276068)^2)^{1/2}$$

Step V: The calculation of relative closeness to ideal solution is represented in Table 17.9.

For instance, the calculation of C_1^* is done as

$$C_1^* = \frac{(0.017109, 0.017852, 0.020563)}{(0.017109, 0.017852, 0.020563) + (0.007502, 0.009096, 0.010586)}$$

$$= (0.54926, 0.66247, 0.83552)$$

Table 17.9 The relative closeness to ideal solution.

Relative closeness	Lower	Modal	Upper
C_1^*	0.54926	0.66247	0.83552
C_2^*	0.45637	0.53894	0.69048
C_3^*	0.60513	0.72864	0.93673
C_4^*	0.59518	0.72410	0.95376
C_5^*	0.21507	0.27979	0.33334
C_6^*	0.17352	0.23875	0.26931

Then, the confidence interval of 60% or **a** = 0.6 is taken for the interval performance measure matrix. A-cut over the fuzzy numbers C_1^*, C_2^*,..... C_3^*, is obtained and it is presented in Table 17.10. Then three values 0.3, 0.5, and 0.7 are taken to find a crisp performance matrix. The optimism index X is applied here and it is presented in Table 17.11. Ranks of the policies are generated according to values of X, and it is presented in Table 17.12.

Table 17.10 Interval performance measure.

Relative closeness	Lower bound	Upper bound
C_1^*	0.617186	0.731688
C_2^*	0.505908	0.599552
C_3^*	0.679237	0.811875
C_4^*	0.672533	0.815963
C_5^*	0.253902	0.301210
C_6^*	0.212660	0.250976

Table 17.11 Crisp performance measure.

Relative closeness	$\varkappa$ =0.3	$\varkappa$ =0.5	$\varkappa$ =0.7
C_1^*	0.651536	0.674437	0.697337
C_2^*	0.534001	0.552730	0.571459
C_3^*	0.719028	0.745556	0.772083
C_4^*	0.715562	0.744248	0.772934
C_5^*	0.268094	0.277556	0.287017
C_6^*	0.224155	0.231818	0.239481

Table 17.12 Rank of different policies by fuzzy TOPSIS method.

Policy	Rank for $\varkappa = 03$	Rank for $\varkappa = 0.5$	Rank for $\varkappa = 0.7$
RP1	4	4	4
RP2	3	3	3
RP3	6	6	5
RP4	5	5	6
RP5	2	2	2
RP6	1	1	1

17.9 Conclusion

In this work, TOPSIS approach is applied to the fuzzy project network. The purpose of the approach is to find out the critical path using different criteria. A triangular hesitant fuzzy set is applied to find the critical path of the project network. Shaik *et al.* [14] have proposed a triangular hesitant fuzzy set's distances measure for selecting a critical path in the new TOPSIS method. Triangular hesitant fuzzy set and hesitant fuzzy project planning and scheduling technique are applied in distance measure. TOPSIS method is explained by illustrating an example. The method has shown the performance in determining critical paths when there were different criteria. Due to the imprecise nature of data for CCPM, the crisp data are converted to fuzzy triangular numbers and fuzzy extensions of crisp and fuzzy MCDM methods are employed to find any variation in the result. fuzzy MCDM methods, namely, fuzzy TOPSIS is employed using fuzzy triangular numbers. The final evaluation of this method also gives the values in form of fuzzy triangular numbers. As the fuzzy triangular numbers' ranking cannot be done like the crisp number, out of several methods of ranking fuzzy numbers, $\boldsymbol{a}$-cut technique is used in the analysis for ranking of fuzzy triangular numbers.

Taking the confidence limit to 60%, i.e., taking $a = 0.6$, the fuzzy triangular number is transformed into interval performance measures and then using the optimism index, it is again transformed to crisp number, where the ranking of the policies is done. Surprisingly, it is found that the same two policies RP6 and RP5 have occupied the first and second ranks as in crisps MCDM methods.

Further, it is suggested to go for alternative policy RP6, which is policy no. P3 for maximization of net benefit, agricultural production and labour utilization.

17.10 Scope of Future Study

a. As the area under the study is small, large area where soil quality is not uniform can be undertaken by using multi criteria decision making approach.
b. More numbers of constraints can be considered like availability of fertilizers, availability of capital, downstream requirements and minimum and maximum reservoir storage.
c. More numbers of criteria can be considered.
d. Fuzzy trapezoidal numbers can be used instead of fuzzy triangular numbers.
e. Other methods, like fuzzy AHP, fuzzy WPM, and fuzzy PROMETHEE, can be used for the said case.

References

1. Ahn, B.S., Extreme point-based multi-attribute decision analysis with incomplete information. *Eur. J. Oper. Res.*, 240, 3, 748–755, 2015.
2. Aruldoss, M., Lakshmi, T.M., Venkatesan, V.P., A survey on multi-criteria decision-making methods and its application. *Am. J. Inf. Syst.*, 1, 1, 31–43, 2013.
3. Giusti, E. and Marsili-Libelli, S., A fuzzy decision support system for irrigation and water conservation in agriculture. *Environ. Model. Software*, 63, 73–86, 2015.
4. Goldratt, E.M., *Critical chain*, North River Press, Great Barrington, MA, 1997.
5. Dwivedi, G., Srivastava, R.K., Srivastava, S.K., A generalized fuzzy TOPSIS with improved closeness coefficient. *Exp. Syst. Appl. Int. J.*, 96, 185–195, 2018.
6. Hartigan, J.A., *Clustering algorithms*, John Wiley & Sons, New York, 1975.
7. Hobbs, B.F., Discussion of multi-objective approaches to river basin planning. *J. Water Res. Plann. Manage. ASCF*, 110, 1, 123–124, 1984.
8. Hwang, C.L. and Yoon, K., *Multi attribute decision making method and application-a state-of-the-art survey*, Springer Verlag, New York, 1981.
9. Jain, A.K. and Dubes, R.C., *Algorithms for clustering data*, Prentice Hall, Englewood Cliffs, New Jersey, 1988.

10. Mainuddin, M. *et al.*, Optimal crop planning model for an existing ground water irrigation project in Thailand. *Agric. Water Manage.*, 33, 43–62, 1997.
11. Garcia, M.S. and Lamata, M.T., A modification of the index of Liou and Wang for ranking fuzzy number. *Int. J. Uncertain. Fuzziness Knowl.-Based Syst.*, 15, 04, 411–424, 2007.
12. Morse, J.N., Reducing the size of the nondominated set: Pruning by clustering. *Comput. Oper. Res.*, 7, 55–66, 1980.
13. Qazvini, Z.E., Haji, A., Mina, H., A fuzzy solution approach to supplier selection and order allocation in green supply chain considering the location-routing problem. *Scientia Iranica, 28*, 446–464, 2021.
14. Shaik, R., Pradha Saradhi, B., Ramesh, H., Ravi Shankar, N., Hesitant fuzzy project planning and scheduling using critical path technique. *Turk. J. Comput. Math. Educ.*, 12, 6, 5272–5286, 2021.

Index

Crisp MCDM methods, 324, 325
Critical chain project management (CCPM). *see* Fuzzy MCDM on CCPM for decision making

CSV form, 55, 58

Also of Interest

Check out these published and forthcoming titles in "Sustainable Computing and Optimization" series from Scrivener Publishing

Optimization Techniques in Engineering
Advances and Applications
Edited by Anita Khosla, Prasenjit Chatterjee, Ikbal Ali and Dheeraj Joshi
Forthcoming 2022. ISBN 978-1-119-90627-8

Computational Intelligence in Sustainable Reliability Engineering
Edited by S. C. Malik, Deepak Sinwar, Ashish Kumar, S. R. Gadde, Prasenjit Chatterjee and Bui Thanh Hung
Forthcoming 2022. ISBN 978-1-119-86501-8

Decision Support Systems for Smart City Applications
Edited by Loveleen Gaur, Vernika Agarwal and Prasenjit Chatterjee
Forthcoming 2022. ISBN 978-1-119-89643-2

Fuzzy Computing in Data Science
Applications and Challenges
Editors: Sachi Nandan Mohanty, Prasenjit Chatterjee and Bui Thanh Hung
Published 2022. ISBN: 978-1-119-86492-9

Handbook of Intelligent Computing and Optimization for Sustainable Development
Edited by Mukhdeep Singh Manshahia, Valeriy Kharchenko, Elias Munapo, J. Joshua Thomas and Pandian Vasant
Published 2022. ISBN 978-1-119-79182-9

Machine Learning Algorithms and Applications
Edited by Mettu Srinivas, G. Sucharitha and Anjanna Matta
Published 2021. ISBN 978-1-119-76885-2

Smart Sustainable Intelligent Systems
Edited by Namita Gupta, Prasenjit Chatterjee and Tanupriya Choudhury
Published 2021. ISBN 978-1-119-75058-1

Printed and bound by CPI Group (UK) Ltd, Croydon, CR0 4YY

17/09/2026

14964252-0006